George Henry Rohé

A Text-Book of Hygiene

A Comprehensive Treatise on the Principles and Practice of Preventive Medicine

from an American Standpoint

George Henry Rohé

A Text-Book of Hygiene
A Comprehensive Treatise on the Principles and Practice of Preventive Medicine from an American Standpoint

ISBN/EAN: 9783337778903

Printed in Europe, USA, Canada, Australia, Japan

Cover: Foto ©berggeist007 / pixelio.de

More available books at **www.hansebooks.com**

OF

HYGIENE

*A Comprehensive Treatise on the Principles and
Practice of Preventive Medicine from an
American Standpoint.*

BY

GEORGE H. ROHÉ, M. D.,

PROF. OF HYGIENE, COLLEGE OF PHYSICIANS AND SURGEONS, BALTIMORE; MEMBER OF THE
AMERICAN PUBLIC HEALTH ASSOCIATION; OF THE AMERICAN DERMATOLOGICAL
ASSOCIATION; OF THE MEDICAL AND CHIRURGICAL FACULTY OF MARYLAND;
CORRESPONDING MEMBER OF THE NEW ORLEANS ACADEMY OF
SCIENCES. ETC.

BALTIMORE :
THOMAS & EVANS
1885.

TO

Henry Ingersoll Bowditch, A. M., M. D.,

THE PIONEER

IN THE FIELD OF

PREVENTIVE MEDICINE

IN

AMERICA.

PREFACE.

THE aim of the author in writing this book has been to place in the hands of the American student, practitioner, and sanitary officer, a trustworthy guide to the principles and practice of preventive medicine.

He has endeavored to gather within its covers the essential facts upon which the art of preserving health is based, and to present these to the reader in clear and easily understood language.

The author cannot flatter himself that much in the volume is new. He hopes nothing in it is untrue.

TABLE OF CONTENTS.

.

CORRECTION:—On page 169, 'sulphuric acid gas' should read *sulphurous acid gas*.

TEXT-BOOK OF HYGIENE.

CHAPTER I.

EXACT investigation into the influence of the atmosphere upon health is yet in its infancy. Enough has been learned, however, to show that changes in the composition of the air, in its density, its temperature, its humidity, its rate and direction of motion, and possibly its electrical or magnetic conditions, influence in various ways the health of the individual. It is only very recently that any scientific attempts have been made to trace the bearing of atmospheric changes upon health. The observations already recorded indicate that a thorough study of meteorological phenomena in connexion with the origin and progress of certain diseases, is a promising field of labor for the educated sanitarian. The meteorological observations which have been gathered by the United States Signal Service during the past thirteen years, already form such a large and tolerably complete and well-arranged body of facts, that reasonably accurate deductions can even now be made. Heretofore, in studying the sanitary relations of the atmosphere, both in this country and abroad, the attention of observers has been riveted almost exclusively upon the changes in its composition occurring within certain limited areas. It is, perhaps, equally important to study this universally diffused and necessary condition of vital activity in its broader and more general relations. It will be shown in the course of the present work, that the meteorological features of countries, or of seasons, or even the daily atmospheric changes, exercise an important influence upon life and health. In order to fully appreciate these relations it will be necessary to first give a brief summary of the facts and laws of meteorology.

THE COMPOSITION AND PHYSICAL CONDITIONS OF THE ATMOSPHERE.

Atmospheric air is a mixture of four-fifths of nitrogen and one-fifth of oxygen ; more accurately 79.00 of the former, to 20.96 of the latter. In addition, there is constantly present a modicum of carbonic acid, usually about .04 per cent., (3 to 5 parts in 10,000), and a variable proportion of vapor of water.

These proportions are maintained, with very little change, at different heights. At first thought, it would seem that carbonic acid, being much heavier than the other constituents of air, would accumulate in the lower regions of the atmosphere, but in obedience to the law of diffusion, the intermingling of the component gases is perfect, and the proportion of carbonic acid in the atmosphere is quite as great on mountain-tops as in the deepest valleys.

The proportion of nitrogen in atmospheric air is generally uniform, while that of oxygen varies, depending to a great extent upon the amount of carbonic acid present. Hence an increase in the amount of the latter constituent is usually accompanied by a diminution of oxygen, inasmuch as the formation of carbonic acid can only take place at the expense of oxygen. The reciprocal activities of animal and vegetable life are beautifully illustrated by these relations between the oxygen and carbonic acid in the air. In the processes of combustion and oxidation, oxygen is withdrawn from the atmosphere and combines with carbon, forming carbonic acid. During vegetable growth, on the other hand, carbonic acid is withdrawn from the air by the leaves of plants, and decomposed into its elements, carbon and oxygen. The carbon is used in building up the plant, while the liberated oxygen is restored to the atmosphere. The animal consumes oxygen and gives out carbonic acid ; the plant resolves this compound into its constituent elements and gives back the oxygen to the air again.

The atmosphere extends upward from the surface of the earth to an indefinite distance. The limit has been variously placed at from forty-five miles to twenty-five thousand miles. In obedience to the law of gravity, this mass of air everywhere presses directly downward—toward the earth's centre—with a force equal to its weight. If a column of this air be balanced by a column, or mass of any other matter—the columns being of the same diameter—we have a relative measure of the weight of the atmosphere. The instrument with which the weight, or downward pressure of the air is measured, is called a barometer. The atmosphere, at the sea-level, presses downward with a force equal to the pressure of a column of mercury thirty inches high. Hence, the barometric pressure at sea-level is said to be thirty inches. If the barometer be carried to the summit of a mountain one thousand feet above the level of the sea, or taken to the same altitude in a balloon, the mercury in the barometer tube will fall about one inch. This inch of the mercurial column represents the weight of the one thousand feet of air now below the barometer, and consequently not measured or balanced by it.*

Upon ascending from the sea-level, it is found also that the air, being less pressed upon by that which is still above it, becomes more rarefied and lighter ; its tension, as it is termed, is less. Hence, for the second thousand feet of ascent above the sea, the mercury will

---

* The figures here given are not absolute, but merely approximate. The limits of this work do not allow a full discussion of the meteorological elements modifying the pressure of the atmosphere at sea-level.

fall a less distance in the tube, the weight removed not being so great as in the first thousand feet.

Variations in temperature and humidity of the air influence the tension of the atmosphere in a marked degree, and affect the height of the barometric column. In fact, most of the changes of atmospheric pressure at the surface of the earth, are directly due to changes in temperature and humidity. Increase of temperature diminishes the density of the air. Hence, when the temperature rises, the pressure decreases. The proportion of moisture (aqueous vapor), if increased, likewise causes a diminution in pressure.

The warmth of the air is primarily derived from the sun. On a clear day about one-fourth of the heat of the sun's rays is given off directly to the air during the passage of the heat-rays to the earth. Of the remaining three fourths, part is reflected from the earth, while the larger portion is first absorbed by the earth and then given off by radiation and convection to the superincumbent air.

The air is always warmer near the earth's surface on a clear, sun-shiny day; for, as soon as the earth gets warmer than the air immediately above it, the excess of heat is given off to the latter by convection and radiation. On ascending from the surface of the earth the temperature decreases, and on the summit of a high mountain, the air is always colder than at its base.

Prof. Tyndall has shown that dry air does not absorb heat. For this reason, the sun's rays strike the earth with much greater intensity on a very dry than on a moist day, while on the latter a larger proportion of the heat rays is intercepted before they reach the earth.

Air at different temperatures is capable of absorbing different amounts of aqueous vapor. Thus, air at a temperature of 40° will require a much smaller amount of vapor to produce saturation than air at a temperature of 80°. For this reason air which appears 'damp' at the former temperature, both to the bodily sensations and to appropriate instruments, would be considered as 'dry' at the latter temperature, although the actual amount of vapor present, or absolute humidity, is the same in both cases.* In meteorological observations for sanitary purposes, the relative humidity is the condition deserving especially careful study.

The motion of the air—wind—is caused by differences in pressure; the latter being due to differences in temperature and humidity. A mass of air traversing a large body of water absorbs vapor, unless already saturated, and becomes moist; if it passes over a wide tract of dry land it loses moisture and becomes dry. Therefore in the eastern portion of the American continent, an easterly or southerly wind which comes from over large bodies of water, and which is usually warm, and thus capable of holding a large quantity of water in a state of vapor, is always moist. On the other hand, a northerly or westerly wind, coming over a large extent of dry land and from a colder region, is nearly always a dry wind. On the Pacific coast these conditions are reversed; there a westerly wind is a moist wind, while an easterly wind is dry. It is probable that the direction and rate of motion of air-currents have considerable influence upon the origin or intensification of certain diseases.

The electrical and magnetic conditions of the atmosphere have been as yet studied to little advantage. It is only known that atmospheric electricity is in most cases positive, and that its intensity increases

---

* By 'absolute humidity' is meant the total amount of vapor present in a certain mass of air. By the term 'relative humidity' meteorologists designate the proportion of vapor present at certain temperatures, compared with full saturation of the air with vapor, which is reckoned 100. Thus, air which is saturated, or whose relative humidity is 100 at 40°, would have a relative humidity of only 24, if the temperature were raised to 80°, because in the latter case the capacity of the air for aqueous vapor is increased. Relative humidity is always designated in percentages; absolute humidity in grains per cubic foot.

with condensation of vapor. There seems to be no
doubt that the varying states of atmospheric electricity
are closely connected with evaporation and condensation. There is reason to believe that a fuller knowledge on these topics will yield most important results
to the student of hygiene.

The presence of ozone in the atmosphere has a very
decided interest to the sanitarian. This modified form
of oxygen which was discovered by Schœnbein in 1839,
is formed wherever there is oxidation without a very
high temperature.* What functions ozone performs in
the air except as an oxidizing agent, is not known.
An impression prevails largely that it has an important bearing upon the causation of disease, but too little
is known upon the subject to form the basis of a rational
theory.

The presence of ozone is detected by test paper, the
best being Schœnbein's, prepared as follows: Add ten
parts of starch to two hundred parts of pure water;
heat it until the starch gelatinizes, and then dissolve
one part of potassium iodide in the mixture. This
paste is then spread on unsized paper and rapidly dried
without exposure to sunlight. The paper is then kept
in a dry, dark place, until wanted for use.†

INFLUENCE OF CHANGES OF ATMOSPHERIC PRESSURE ON
HEALTH.

The effects of a considerable diminution of pressure are familiar to every one in the 'mountain sickness'
which attacks most persons on ascending high mountains. M. Bert has shown experimentally that similar
effects can be produced in an air-tight chamber by diminishing the pressure.‡ The symptoms produced under a

---

* R. C. KEDZIE : On Ozone. Third Annual Report of the Michigan State Board
of Health, 1875, p. 137.—On Ozone and Atmospheric Electricity, see an instructive
paper by the late Dr. GEORGE M. BEARD, in Popular Science Monthly, Vol. IV., p, 456.
    † See KEDZIE, above quoted, p. 143.
    ‡ Popular Science Monthly, V., p. 379.

pressure equivalent to an altitude of from 13,448 feet,
to 16,728 feet, were a feeling of heaviness, nausea, ocular fatigue, rapidity of pulse and convulsive trembling
on slight exertion, and a sensation of languor and general indifference to the surroundings of the individual.

M. Lortet, who has left on record his experiences
in the higher Alps, says that the symptoms noticed on
ascending to high altitudes are labored respiration,
increased rapidity of pulse, depression of temperature,
(as much as 4°–7° C). The normal temperature was
restored, however, after a brief rest.* Still more severe
symptoms have been noticed on ascending high mountains in South America and Asia. Aeronauts have lost
consciousness, and in several instances life, on rapidly
ascending to great altitudes.† According to the observations of the brothers Schlagintweit, distinguished
explorers of the highlands of Asia, the effects of diminished pressure upon the human organism are: 'headache, difficulty of respiration, and affections of the
lungs,—the latter even proceeding so far as to occasion blood-spitting, want of appetite and even nausea, muscular weakness, and a general depression and
lowness of spirits. All these symptoms, however, disappear in a healthy man almost simultaneously with his
return to lower regions.' A singular observation was
made by these travelers on the effect of motion of the
air upon the symptoms described. They say: 'The
effects here mentioned were not sensibly increased by
cold, but the wind had a most decided influence for the
worse upon the feelings. . . . When occupied with
observations, we took very little, if any bodily exercise,
sometimes for thirty-six hours; it would frequently
occur nevertheless, even in heights not reaching 17,000
feet, that an afternoon or evening wind would make us

---

* Real-Encyclopædie d. ges. Heilk. V., p. 529.

† MM. Sivel and Croce-Spinelli, two æronauts, lost their lives in this manner
during an ascent from Paris, in April, 1875.

all so sick as to take away every inclination for food. No dinner was cooked; the next morning when the wind had subsided, the appetite was better.

'The effects of diminished pressure are considerably aggravated by fatigue. It is surprising to what degree it is possible for exhaustion to supervene; even the act of speaking is felt to be a labor, and one gets as careless of comfort as of danger. Many a time our people,—those who ought to have served us as guides —would throw themselves down upon the snow, declaring they would rather die upon the spot than proceed a step further.'*

These symptoms disappear when persons are exposed to these conditions for a prolonged time. Thus, in the Andes there are places thirteen and fourteen thousand feet above sea-level, which are permanently inhabited; and in the Himalayas there are villages at a height of over 15,000 feet constantly occupied. In this country, Pike's Peak, 14,150 feet above the sea, has been occupied since 1873 by observers of the signal service. The men seem to become acclimated, as it were, and suffer little or no inconvenience from the diminished pressure, after a time.

The effects of increased pressure of the atmosphere are especially manifested in divers, and in workmen in deep mines and tunnels. The symptoms are decrease of respiration, diminution in frequency of the pulse, ringing in the ears, earache, headache, mental depression and sometimes actual deafness. 'When the workmen leave the compressed air, they are said to suffer from hemorrhages and occasional nervous affections, which may be from cerebral or spinal hemorrhages.'† The workmen, however, soon become habituated to these

---

* Results of a Scientific Mission to India and High Asia. By HERMANN, ADOLPHE and ROBERT DE SCHLAGINTWEIT. Vol. II., pp. 484-5.

† PARKES' Practical Hygiene, 6th ed., New York, Vol. II., p. 92.

changes and then suffer little inconvenience upon exposure to the increased pressure.

INFLUENCE OF CHANGES OF TEMPERATURE ON HEALTH.

Many of the derangements of health ascribed to
high temperature are to a considerable degree due to
other factors, prominent among which are high humidity, intemperance, overwork and overcrowding. There
can be little doubt, however, that the importance of the
high temperature itself can hardly be overrated. It has
been generally accepted heretofore that a high temperature together with a high relative humidity is most
likely to be followed by sunstroke. A careful comparison in a series of deaths from sunstroke in the city of
Cincinnati in the summer of 1881, shows, however, conclusively, that a very high mean temperature with a
low relative humidity is more liable to be followed by
sunstroke than the high temperature when accompanied
by a high humidity. The same series of observations
also shows that the number of deaths was greater on
clear days than on cloudy or partly cloudy days.* A
corroboration of this result is found in the fact that
sunstrokes very rarely occur on shipboard, at sea,
where the relative humidity is always high.

Diarrhœal diseases, both of adults and children,
are much more frequent during hot than during cold
weather, (and in hot than in cold climates), but it is
probable that other factors aid in the production of
these diseases beside the high temperature.

Certain epidemic diseases are likewise more frequent in, or exclusively confined to, hot climates.
These are cholera, yellow fever and epidemic dysentery.
Elephantiasis and the prevalence of certain skin diseases seem also to have some connexion with a constantly high external temperature. The intimate rela-

* A. J. MILES: The Sunstroke Epidemic of Cincinnati, O., during the Summer
of 1881. Public Health, Vol. VII., p. 293-304.

tion between cause and effect is not clearly understood, although the belief is current that the origin and spread of such diseases depend upon the development of various parasitic organisms.

Extreme low temperature, as observed in the arctic regions, seems to produce a progressive deterioration of the blood, (anemia), in consequence of which most natives of temperate regions who are compelled to remain in the far north longer than two winters, succumb to various hemic diseases, scurvy being the most prominent. It is not improbable, however, that the dietary furnished is responsible for a large share of the evil effects ascribed to cold. The absence of sunlight for a considerable part of the winter season, may also have much to do with the bad influences for which the low temperature is held responsible.

Among the acute effects of great cold, frost-bite is the most frequent, as well as the most serious. Loss of portions of the nose, or ears, or even of entire members are not infrequent results of frost-bite.

In the arctic regions, one of the most annoying affections which the traveler has to contend against, is snow-blindness, a severe ophthalmia produced by the glare of the snow. Neutral tinted glass goggles should be worn as a preventive.*

HUMIDITY OF THE ATMOSPHERE AS CONNECTED WITH CHANGES IN HEALTH.

The propagation of certain acute infectious diseases is believed to be due to a high relative humidity. There can be no longer any doubt that a very humid soil and air, especially if connected with a variable temperature, are almost constant factors in the produc-

---

* See PAYER's Narrative of the Austrian Arctic Voyage of 1872-74, p. 250-3 and 317, for an account of the effects of cold on the organism, and on the best prophylactic measures to be adopted. The Report of the Surgeon General of the U. S. Navy for 1880, also contains (pp. 330-8) a valuable memorandum by Surgeon-General PHILIP S. WALES, on Arctic Hygiene.

tion of pulmonary phthisis. Recent experience in this country and abroad has shown that the high plateaus and mountains, far inland, where the soil is dry and the relative humidity of the air low, are the best resorts for consumptives.

Of the effects of excessively dry air on health, little definite is known. It seems probable, however, that catarrhal affections of the respiratory mucous membrane are more frequent in a dry than in a humid climate.

With reference to the influence of electrical conditions of the atmosphere upon health, no observations have been made which admit of definite conclusions being reached.*

Mr. Alexander Buchan and Dr. Arthur Mitchell have analyzed the influence of the weather and season upon the causation of disease, or rather upon the mortality from various diseases.† Taking the records of the city of New York from 1871 to 1877, it appears that the maximum number of deaths from small-pox occurred in May, the minimum in September. From measles there were two annual maxima and minima, the greater in July and September, and the smaller in February and April. From scarlet fever the maximum was in April, the minimum in September. From typhoid fever the maximum was from August to November, the minimum almost equally distributed throughout the rest of the year. From diarrhœa, the maximum in July and August, the minimum from Decem-

---

* DR. S. WEIR MITCHELL has shown (from the record of the case of Capt. Catlin, U. S. A., see American Journal Med. Sci., April, 1877, and N. Y. Med. Jour., Aug. 25, and Sept. 1, 1883), that attacks of neuralgia, in this case at all events, accompanied the progress of storms across the continent. Also, that the periods of maximum pain occurred with a high but falling barometer and increasing absolute humidity. There seems also to be some relation in this case between the maximum pain and the maximum magnetic force as shown by the declinometer. DR. MITCHELL's papers are among the most valuable positive contributions to hygienic meteorology and deserve careful study.

† Journal Scottish Meteorological Society, 1875-8. Abstract in RICHARDSON's Preventive Medicine, Phila., 1884, p. 533, et seq.

ber to March. From diphtheria the maximum in December, the minimum in August.* From whooping-cough, maximum in September and February, minimum in November and June. For croup, the curves agree pretty closely with the diphtheria curves. From phthisis, the maximum in March, minimum in June. From suicide, curiously the greater number occurs in May, the smallest in February. This is contrary to the usual supposition that gloomy weather predisposes to suicide. In Philadelphia, the results of an examination of the statistics of suicide for ten years are almost exactly similar. Out of 636 cases of suicide, 78 occurred in May, 71 in August, 57 in December, 54 each in October, July and April, 52 in June, 49 in November, 44 each in December and February, 43 in March, and 36 in January.† Dr. Lee is led to believe 'that a low barometric pressure, accompanied by a high thermometric registry, with sudden fluctuations from a low to a high temperature, together with much moisture and prevailing south-west winds, might somewhat account for the frequency of self-murder in the spring and summer months.'

THE SANITARY RELATIONS OF CHANGES IN COMPOSITION AND OF IMPURITIES IN THE AIR.

The average proportion of carbonic acid in the atmosphere is from 3 to 5 parts in 10,000. Pettenkofer ‡ places the maximum limit of carbonic acid allowable in the air of dwellings, at 7 parts in 10,000. It is probable that this limit is very frequently exceeded without serious consequences to health, if the air is not at the same time polluted by organic impurities, the

---

* See paper on the relation of weather to mortality from diphtheria in Baltimore. By RICHARD HENRY THOMAS, in Trans. Med. and Chir. Faculty of Maryland, 1883.

† JOHN G. LEE: Suicide in the City and County of Philadelphia during a decade, 1872 to 1881 inclusive. Trans. Am. Med. Asso., Vol. 33, p. 425.

‡ Quoted in BUCK's Hygiene and Public Health, Vol. I., p. 615.

products of respiration. Prof. William Ripley Nichols, found the air in a school-room in Boston to contain eight times the normal proportion of carbonic acid, while Pettenkofer found, also in a school-room, after the same had been occupied two hours, eighteen times the normal proportion, or 72 parts in 10,000.* While such an excess of this poisonous gas must unquestionably have an unfavorable influence upon health, it is probable that the most serious effects are due to the coincident diminution of oxygen and the pollution of the air by the products of respiration which necessarily take place during respiration. Carbonic acid alone may be present in the air to a much greater extent than above mentioned, without causing any appreciable inconvenience. In the air of soda-water manufactories there is frequently as large a proportion as two per cent. of this gas present without producing any ill effects upon those breathing such an atmosphere.

The amount of carbonic acid in the atmosphere is greatest at night. It is also greater very near the ground than at a distance of several feet above it. As carbonic acid is absorbed by the leaves of plants during the day-time, but given off at night, the difference may partly be thus accounted for. According to Fodor,† the source of a large proportion of the carbonic acid in the air, is the decomposition going on in the soil. This accounts for the larger percentage of carbonic acid near the ground. This would also explain the variation of the proportion of carbonic acid in the air under different meteorological conditions. For example, it is found that during rainy weather the carbonic acid in the air is diminished. This is accounted for partly by the absorption of the carbonic acid by the saturated ground,

---

* See table in BUCK's Hygiene and Public Health, Vol. I., p. 612.

† FODOR: Hygienische Untersuchungen ueber Luft, Boden und Wasser, Braunschweig, 1882, 2te Abth.

while at the same time the porosity of the soil is diminished and the escape of the ground air prevented.

Mr. R. Angus Smith made a number of experiments upon himself to determine the effects of an atmosphere gradually becoming charged with the products of respiration and perspiration. His experiments
were conducted in a leaden chamber holding 170 cubic
feet of air. This air was not changed during the experiment. After remaining for an hour in this chamber,
an unpleasant odor of organic matter was perceptible
on moving about. The air when agitated felt soft, owing doubtless to the excess of moisture contained in it.
The air soon became very foul, and although not producing any discomfort, the experimenter states that
escape from it produced a feeling of extreme pleasure,
like 'that which one has when walking home on a fine
evening after leaving a room which has been crowded.'*

Hammond† confined a mouse in a large jar in which
were suspended several large sponges saturated with
baryta water, to remove the carbonic acid as rapidly
as formed. Fresh air was supplied as fast as required.
The aqueous vapor exhaled was absorbed by chloride
of calcium. The mouse died in 45 minutes, evidently
from the effect of the organic matter in the air of the
jar. The presence of this organic matter was demonstrated by passing the air through a solution of potassium permanganate.

The horrible story of the 'black hole' of Calcutta
is familiar to every one. Of 146 prisoners confined in
a dark cell at night, 23 were found alive in the morning. Among the survivors a fatal form of typhus fever
broke out, which carried off nearly all of them. After
the battle of Austerlitz, 300 prisoners were crowded in
a prison; 260 died in a short time from inhaling the

---

* Air and Rain, p. 138.

† A Treatise on Hygiene, with special reference to the Military Service. By Wm.
A. Hammond, M. D., Surgeon-General U. S. Army. Philadelphia, 1863, p. 170.

poisoned air. Numerous other similar examples of the effects of polluted air are recorded.

Usually the effects of foul air are not so sudden and striking. In most instances, especially where the pollution has not reached a high degree, there simply results a general deficiency of nutrition, which manifests itsef in anemia, loss of vigor of body and mind, and a gradual diminution of resistance to disease.

It seems to be beyond question that persons who are constantly compelled to inhale impure air, especially if combined with an improper position of the body, or lack of sufficient or appropriate food, furnish a very large percentage of chronic pulmonary affections. Phthisical patients, in the overwhelming majority of cases, are drawn from the classes whose occupations keep them confined in close rooms. Want of exercise, and of good food, doubtless aid in the development of the lung disease. Formerly, when less attention was paid to the proper construction and ventilation of barracks and prisons, the mortality from phthisis among soldiers and criminals was much greater than it is now. In animals kept closely confined, the same disease claims a large share in the mortality.

Near the end of the last century, over one-third of the infants born in the old Dublin Lying-in Hospital, died of epidemic disease. After the adoption of an improved system of ventilation, the mortality fell to about one-tenth of what it had previously been. To illustrate the effect of similar conditions upon the health of domestic animals, the following instance is cited: Upward of thirty years ago, a severe epidemic of influenza in horses appeared in Boston. At the instigation of Dr. H. I. Bowditch, every stable in the city was inspected and classified, as 'excellent,' 'imperfect,' or 'wholly unfit,' in respect to warmth, dryness, light, ventilation and cleanliness. It was found that in the first class fewer horses were attacked and the disease was milder, while in the third class every horse was attacked, and the more severe and fatal cases occurred.

Carbonic oxide gas is a very dangerous impurity often present in the air of living-rooms. Being an ingredient of illuminating gas, as well as the so-called coal-gas, which so frequently escapes from stoves and furnaces, its dangerous character becomes apparent. Many persons die every year in this country from the inhalation of illuminating gas. People unacquainted with the mechanism of the gas-fixtures, frequently blow out the light instead of cutting off the supply of gas by turning the stop-cock. It is also a prevailing custom to keep the light burning 'low' during the night. Any considerable variation of pressure in the pipes, or sudden draught, may put out the light and permit the gas to escape into the room with fatal effect. Leaks in the pipes or fixtures may have the same results.

Coal, coke, or charcoal fires may produce serious or fatal poisoning, if the gas, which contains a large proportion of carbonic oxide, is permitted to escape into the room. In certain parts of Europe, notably in France, the inhalation of the fumes of a charcoal fire is a favorite method of committing suicide.

Sulphuretted and carburetted hydrogen are not infrequently present in the air, especially about cesspools, and in mines and certain manufacturing establishments. Sulphuretted hydrogen is generally considered to be a violent poison, but there is no evidence that it is so unless oxygen is excluded.

Carburetted hydrogen is the so-called 'fire-damp' of mines, which is so often the cause of fatal explosions. Its inhalation does not seem to be especially noxious. It will be more fully referred to in a succeeding chapter.

Variations in the proportion of ammonia present in the air are frequent. Its presence is an indication of organic decomposition in the vicinity, but nothing is known of the influence of the gas itself upon health, in the proportion in which it is ever found in the atmosphere.

SEWER AIR.

Sewer air, or sewer gas, as it is often improperly called, is a variable mixture of a number of gases, vapors, atmospheric air and solid particles, and is derived from the decomposition of the animal and vegetable contents of sewers. A number of analyses by different chemists have shown that the composition of sewer air is extremely variable. The most important components, in addition to the constituents of atmospheric air, are: carbonic acid, ammonia, sulphuretted hydrogen, and a number of volatile organic compounds which give to sewage its peculiar odor, but which are present in such small quantity as to prevent accurate determination by chemical means. Sewer air may also contain particulate bodies, bacteria and other microscopic organisms which are supposed by many to be the active causes of infectious diseases.

When the contents of sewers remain in these receptacles or conduits long enough to undergo decomposition, sewer air is always present. It is believed by some physicians and sanitarians that sewer air is the direct cause of such diseases as typhoid fever, scarlet fever, diphtheria and cholera, while others hold the view that the sewer air is merely a favorable breeding place for the germs of these diseases, and that it thus merely acts as a medium in which the infective agent grows, reproduces itself, and is conveyed from place to place. There is no absolutely reliable evidence in favor of either of these doctrines.

It is hardly open to question, however, that the continual breathing of air polluted by emanations from sewers often produces more or less serious derangements of health. Diarrhœa, and other intestinal affections, mild cases of continued fever, and even cases of undoubted typhoid fever have been so frequently noted in connexion with defective sewerage, and the escape of sewer air into inhabited rooms, that doubt upon this point is hardly justifiable. With regard to typhoid fever, however, it is probable that the sewage in these cases contained the particular virus which causes this disease.

The effluvia from cemeteries, knackeries, and other places where the bodies of animals are undergoing decomposition are, it is believed with good reason, deleterious in their effect upon health.

Prof. Tyndall has shown* that even the apparently clearest air is, when in motion, constantly filled with innumerable particles of dust, which are believed by many to give rise to various forms of disease. The presence of these particles can be easily demonstrated by means of the electric light. Everyone has observed these minute particles in a bright ray of sunlight. Under ordinary conditions these particles of dust would of course give rise to no trouble, but if intermingled with these dust specks there were disease germs— whether these germs be considered as living organisms, or as particles of dead tissue from the body, then manifestly the inhalation of such 'dust' would be dangerous.

The quantity of dust found in the air of cities is much greater than in the country. Tissandier found that in Paris the percentage of dust was eight to twelve times greater than in the open country. One-fourth to nearly one-half of this atmospheric dust is organic, either animal or vegetable. Very recent observations have shown that in Paris the air contains nine or ten

---

* Essays on the Floating Matter of the Air. New York, 1882.

times as many bacteria in a given volume as the air at the observatory of Montsouris, just without the city.

TESTS FOR IMPURITIES IN THE AIR.

The sense of smell will indicate the presence of sulphuretted hydrogen, or of volatile organic matter. Chemical tests and the microscope will, however, be necessary to determine the presence of carbonic acid, carbonic oxide, or suspended particulate matter in the air.

In order to detect the presence of carbonic acid, advantage is taken of the affinity of the acid for certain alkalies with which it forms insoluble compounds. If a stream of carbonic acid gas is passed through lime or baryta water, an insoluble carbonate of lime or baryta is instantly formed, and produces a milky precipitate in the water. If, instead of passing a stream of the gas through the liquid, the latter be agitated with air containing carbonic acid, a similar precipitate is produced. The most exact method of determining the amount of carbonic acid in the air is that known as Pettenkofer's,* but it is somewhat complicated. A readier method has been devised by Mr. Angus Smith, and is termed the minimetric test.† A series of six wide-mouthed bottles, having a capacity respectively of 450, 350, 300, 250, 200 and 150 cubic centimeters‡ is fitted with clean, tightly-fitting corks. The bottles are made perfectly clean and dry, and 15 cu. cm., (3¾ drams) of clear, fresh lime or baryta water put into the smallest, the cork replaced and the bottle well shaken. If the water becomes turbid there is at least .16 per cent (16 parts per ten thousand) of carbonic acid in the air treated. If only the water in the largest bottle becomes

---

* Nowak, Lehrbuch der Hygiene, p. 149.

† Nowak, op. cit., p. 152.

‡ The equivalents in English measures are: 14 oz., 11 oz., 9½ oz., 8 oz., 6¼ oz., and 4¾ oz.

cloudy, the proportion of carbonic acid is probably less than 5 parts in 10,000. For the intermediate series of bottles, the amounts of carbonic acid necessary to produce cloudiness are respectively: for 200 cu. cm. of air, 12 parts in 10,000; for 250 cu. cm., 10 parts; for 300 cu. cm., 8 parts; and for 350 cu. cm., 7 parts per 10,000. If therefore a cloudiness is produced with any of the bottles except the largest, the amount of carbonic acid present exceeds the standard allowable in pure air. The test should be frequently made, in order to acquire familiarity with its use. The same quantity of the test liquid is of course used in each bottle.

Carbonic oxide is detected by its reaction with chloride of palladium, which gives a black color when brought in contact with the gas. If a strip of linen, or blotting paper be moistened with a solution of the palladium chloride (gr. j to ℥ j), and suspended in air containing carbonic oxide, the black color will be developed. The suspected air may also be passed through a solution of sodio-chloride of palladium, when the liquid will turn black, if carbonic oxide be present.

The percentage of organic impurity in the air of an occupied room (products of respiration, etc.) is difficult to ascertain directly. Pettenkofer has found, however, that the proportion of carbonic acid present is indirectly a measure of the organic impurity from respiration.* As the determination of the carbonic acid is easy by the minimetric method of Angus Smith, the extent to which the air is polluted by respiratory impurities is readily ascertained.

The presence of organic and other suspended impur-

---

* Recent observations in this country (see Annual Reports of the Surgeon-General of the Navy for 1879, p. 45 and 46, and the same for 1880, p. 31-34,) seem to throw some doubt upon the entire reliability of this method of determining the amount of organic matter in the air examined. Prof. IRA REMSEN (Rept. Nat'l Board of Health, 1879, p. 77, and 1880, p. 308, et seq.,) has shown the insufficiency of the chemical methods at present in use, and points out the difficulties of making reliable and satisfactory determinations of organic matter in the air.

ities can be best demonstrated with a microscope. An objective magnifying upward of 400 linear diameters, and experience in the use of the instrument, will be needed to obtain correct results. By moistening a glass slide with glycerine, and exposing it in the suspected air, a sufficient quantity of the suspended matters may be collected in the course of twenty-four hours to permit some conclusions to be drawn from a microscopic examination.*

A common method of determining the presence or absence of a large quantity of carbonic acid, for example, at the bottom of a well or privy-vault, is to lower a lighted candle to the bottom. If the light is extinguished, the air is considered irrespirable ; but if it continues burning brightly, the air is believed to be sufficiently pure to sustain life. Sulphuretted hydrogen and sulphide of ammonium are sometimes found in privy-vaults, and although they will not extinguish a light, they speedily prove fatal if inhaled in a concentrated form, and to the exclusion of a sufficiency of oxygen. Cases frequently occur where serious or fatal results have ensued from the presence of a dangerous gas, which was thought to be excluded by the burning candle.†

It is advisable in all cases to exhaust the stagnant air in all old wells and privy-vaults, before permitting any one to descend. Perhaps the readiest method of exhausting the vitiated air in such places would be to lower heated stones, masses of hot iron or pails of hot water, to near the bottom, which produce a rarefaction of the air and cause it to ascend. Its place will then be occupied by purer air from without. The rarefac-

---

* Dr. G. M. Sternberg, U. S. A., (Rept. National Board of Health, 1880) gives an account of his investigations into the suspended matters of the air. The question is also considered in a practical manner by Surgeons Kidder and Streets, U. S. N., in Reports of the Surg.-Gen'l of the Navy for 1880 and 1881.

† See a case reported in Philadelphia Medical Times, October 21, 1882.

tion produced by the explosion of gun-powder has also
been made use of with success ; but this has some
objections, because the combustion of powder itself
produces gases which are noxious if breathed in large
quantity.   An animal, such as a cat or dog, should be
first lowered into the suspected well for fifteen or
twenty minutes, in order to determine whether the air
at the bottom is capable of sustaining life, before permitting the workmen to descend.   Similar precautions
should be used in old, unused mines, to prevent fatal
effects from the so-called 'choke damp' which is largely composed of carbonic acid.

VENTILATION.

It is generally accepted that the presence of .07 per
cent. (seven parts in 10,000) of carbonic acid in the air,
indicates the greatest amount of organic impurity
(from respiration or combustion) consistent with the
preservation of health.   Adopting this as the standard
of maximum impurity allowable, 3,000 cubic feet of
fresh air per hour will be needed for each individual
to keep him supplied with pure air.   This is for persons
in a state of health ; in cases of disease, a more rapid
change of air will be necessary to keep that surrounding the patient in a state of purity.

Ventilation is defined by Worcester, as 'the re-
placement of noxious or impure air in an apartment,
mine, or enclosed space, by pure fresh air from without.'
By Dr. Parkes the term is restricted to 'the removal
or dilution, by a supply of pure air, of the pulmonary
and cutaneous exhalations of men, and the products of
combustion of lights in ordinary dwellings, to which
must be added, in hospitals, the additional effluvia
which proceed from the persons and discharges of the
sick.   All other causes of impurity of air ought to be
excluded by cleanliness, proper removal of solid and

liquid excreta, and attention to the conditions surrounding dwellings.'*

A proper system of ventilation must take into consideration the cubic space of the apartment or building to be ventilated, the number of persons ordinarily inhabiting this space, and certain other collateral elements, such as the character of the building to be ventilated, its exposure, necessity for artificial heating, etc.

The amount of cubic space that must be allowed to each individual is determined by the rapidity with which fresh air must be supplied in order to keep that surrounding the individual at the standard of less than .07 per cent. of carbonic acid. For example, in a space of 100 cubic feet, the air must be changed thirty times in an hour, in order to prevent the carbonic acid exceeding the above proportion, that is to say, to allow 3,000 cubic feet of air to pass through that space in the time mentioned. This would create an uncomfortable, if not injurious draught. If the space contained 1,000 cubic feet, the air would need renewal only three times an hour.

A space of 500 cubic feet could be kept supplied with pure air without perceptible movement if all the mechanical arrangements for changing the air were perfect ; but such perfection is rarely attainable, and hence there would be either draughts, or insufficient ventilation in such a small 'initial air space,' as it is termed. The initial air space should, therefore, be not less than 800, or better, 1,000 cubic feet. The air of this space could be changed sufficiently often to keep it at its standard of purity, without creating unnecessary draught. For sick persons this should be doubled. In hospitals, therefore, the cubic air space allowed to each bed should not be less than 1,800 to 2,000 cubic feet.

---

* Manual of Practical Hygiene, 6th edition, New York, Vol. I, p. 157.

The source of the air supplied, must of course, be capable of yielding pure air. It should not be drawn from damp cellars or basements, or from the immediate vicinity of sewers, or drains. Air taken from such places is little better for respiration than that which it replaces in the apartments to be ventilated.

Ventilation may be accomplished either with or without artificial aids. In buildings or rooms, used as habitations, natural ventilation, with, perhaps, the simplest mechanical aids are made use of almost entirely. In large buildings, such as churches, theatres, schools, or in ships and mines, one of the artificial systems must be adopted if efficient ventilation is desired.

Natural ventilation takes place by diffusion, by perflation, and in consequence of inequality of atmospheric pressure. By diffusion, is meant the slow and equable entrance of air from without, and exit from within a room through the walls or ill-made joints, without the influence of wind currents. In an occupied room this is, however, insufficient to keep the air pure, because many of the organic impurities of respired air are molecular, and, therefore, incapable of making their way out of the rooms through the walls.

Perflation means literally, 'blowing through,' and if the direction and force of air-currents could be regulated, this would, with simple mechanical arrangements, be an efficient means of ventilation. However, the uncertainty of the force and direction of the wind make this method of ventilation unreliable, except in warm weather.

Unequal pressure between the air in a room, and that without, is, within certain limits, an efficient means of ventilation, and is usually relied upon in ordinary apartments. When the air in a room is heated above the temperature of the external air, either by a fire, lights or by the presence of a number of persons in the room, it expands and part of it finds its way out through numerous crevices and bad joints found in all buildings. The air which remains, being less dense than the external air, the latter enters the room by various openings, until the equality of pressure is re-established. But as the heating of the enclosed air continues, the process is momentarily repeated and becomes continuous.

Although the impurities of respired air (carbonic acid, organic matters), are heavier than the air itself at the same temperature, it is a familiar fact that the most impure air in an occupied room is always found near the ceiling, the impurities being carried upward with the heated air, and that the pure air from without, being colder, fills the lower part of the room.

If the cold, outside air, were to be admitted at the bottom of the room, and means allowed for the escape of the hot air at the top, the conditions of the old health-maxim, to 'keep the feet warm and the head cool,' would be reversed. This would be no less uncomfortable than unwholesome. In all plans for natural ventilation, therefore, provision must be made to secure a gradual diffusion of the cold, outside air from above, or to have it warmed before it enters the room. With a large chimney as an aspirating shaft,* with flues at the top and bottom of the room, and openings in the walls of the room near the ceiling to admit fresh air, sufficient ventilation can be usually secured in cold weather, in a room not overcrowded.

When a room is heated by a furnace, the fresh air is warmed before it is introduced, and the foul air escapes either through a ventilating shaft, a ventilator

---

* Of course there is really no such thing as a real aspiration, or 'sucking out' of the air through the chimney or so-called 'aspirating shaft.' The upward movement of the air in the shaft is due to its displacement by the colder or denser air entering the room.

in the window, or wall, or through the numerous fissures and other orifices, which defective carpentering always leaves for the benefit of the health of the occupants.

The following rules for the arrangement of a system of natural ventilation. are modified and condensed from Parkes :*

The apertures of entrance and of exit for the air should be placed far enough apart to permit thorough diffusion of the fresh air.

When the air is brought into a room through slits or tubes in the walls near the ceiling, the current should always be deflected upward by an inclined plane, in order to prevent a mass of cold air from descending over the shoulders of the occupants and chilling them.

The air must be taken from a pure source.

The inlet tubes should be short, and so made as to be easily cleansed, otherwise dirt lodges and the air becomes impure.

Inlets should be numerous and small, to allow a proper distribution of the entering air.

Externally, the inlets should be partially protected from the wind to prevent strong draughts ; they should also be provided with valves to regulate the supply of air.

If the air cannot be warmed, the inlets must be near the ceiling ; if it can be heated, it may enter near the floor.

The air may be warmed by passing it through boxes containing hot water, or steam coils; by passing it through chambers around grates or stoves, or heating it in a furnace.

In towns or manufacturing districts, the air should be filtered before allowing it to enter the room. Thin

---

* Manual of Practical Hygiene, 6th ed., New York, Vol. I, p. 177.

flannel or muslin spread over the openings answers very well as filtering material.

Outlets should be placed at the highest point of the room and should be protected from the weather. An opening into the chimney near the ceiling will answer well in many cases.

In one story buildings, ridge-ventilators make the best outlets. The entrance of snow and rain must be prevented by suitable arrangements.

A small space or slit between the horizontal bars of the upper and lower window sash will admit sufficient air in a proper direction in small rooms, even when the window is shut.

A good arrangement is one devised by Dr. W. W. Keen, of Philadelphia.* A piece of stout paper or cloth is tacked across the lower ten or twelve inches of the window frame. Raising the lower sash, allows the entrance of fresh air, which is directed upward by the cloth obstruction.

In all rooms, howsoever ventilated, doors and windows should be often opened to permit a thorough *flushing* of the interior with fresh air.

For large buildings, hospitals, schools, theatres, ships, and mines, two systems of artificial ventilation are in use. One operates by extracting the foul air by means of fans, the other, by forcing in fresh air, allowing the impure air to find its way out as best it may.

In some of the vessels of the U. S. Navy a system of ventilation is in use, for the details of which the author is indebted to Dr. A. L. Gihon, Medical Director, U. S. N. The apparatus in general consists of two mains running the whole length of the vessel, opening into a blower room in which a fan revolves. The revolving fan causes an up-draught exhausting the air from the mains. From these, small pipes run to every state-room,

* HARTSHORNE, H. Our Homes, p. 64.

store-room, passage-way, frame-space and cul-de-sac in the vessel. By opening the frame-spaces* on the upper deck (an original device of Dr. Gihon), fresh air is freely supplied, bathing the whole frame or hull, and passing directly into the bilges, diluting the foul air, and thence being drawn by the aspirating force of the fan into the main ventilator conduits.

[In addition to the works mentioned in the text, the following may be referred to as more fully treating of the subjects considered in this chapter:

FLAMMARION: The Atmosphere.—The Articles on Atmosphere, and Climate in the ENCYCLOPEDIA BRITTANICA, 9th Edition.—Reports of the Chief Signal Officer of the Army.—A paper on Climate and Diseases, by Dr. CLEVELAND ABBE, in Report of National Board of Health for 1880.—Die Canalgase, by Dr. F. RENK, Munchen, 1884.--MORIN: On Warming and Ventilating occupied Buildings; translated in Smithsonian Report for 1873 and 1874.]

* Frame-spaces are the spaces between the timbers or ribs of a vessel.

CHAPTER II.

WATER.

PHYSIOLOGISTS teach that nearly two-thirds of the tissue of the animal body consists of water. Inasmuch as this water is constantly being lost by evaporation from the skin, exhalation by the lungs, and excretion through various organs, it is evident that the loss must be constantly supplied if the functions of life shall be properly performed.

It appears probable that certain diseases are at times spread through the agency of insufficient or impure drinking water. It is therefore a matter of very great importance to have a definite notion of what constitutes a pure and sufficient supply of water, and how best to secure it; to be able to detect its conditions of purity and impurity, and to know how to maintain the former, and avoid the latter. It will be necessary to consider in detail, therefore, the quantity of water required by each individual for the maintenance of health, the sources whence water is obtained, how it shall be collected and stored to the best advantage, the impurities likely to be contained in it, and the methods of keeping it pure, or of purifying it when it has become polluted or vitiated in any manner.

THE QUANTITY OF WATER REQUIRED BY HUMAN BEINGS.

Dr. Parkes, after a number of experiments, concluded that a man of the English middle class, 'who may be taken as a fair type of a cleanly man belonging to a fairly cleanly household' uses about twelve gallons of water per day. This covers all the water needed, including a daily sponge bath. Dr. DeChaumont estimates* that sixteen gallons should be the daily allow-

---

* PARKES' Hygiene, 6th Ed., N. Y., Vol. I., p. 5.

ance. By order of the British War Department, 15
gallons of water are allowed to each soldier daily.
In very many instances, this quantity cannot be furnished, but in such cases there necessarily results some
deficiency in cleanliness. It is probable that among
the poorer classes, especially where a large supply of
water is not convenient, the quantity used is not over
one-fourth of the above estimate.

In estimating the daily supply of water needed in
a community, large or small, other circumstances must
be taken into consideration, in addition to the demands
of the individual. For example, in towns or cities,
allowances must be made for animals, manufacturing
purposes, probable waste, fires, sewerage, etc. In cities
an allowance of 50 gallons daily per head would not be
excessive. In most American cities the supply is much
greater. The present daily supply in Baltimore is estimated at 60 gallons per head,* which could be increased
to three times that quantity if necessary.

A serious problem, affecting, however, the engineer,
rather than the sanitarian, is the prevention of waste
of water in places where the supply is limited. It is
estimated that in Chicago one-half of the water pumped
is wasted through negligence and imperfections in the
supply apparatus, while in St. Louis, the annual cost to
the city of the water that is wasted is placed at $400,000.†
It has been proposed to check this wanton waste by
measuring the quantity of water used by each household by means of a meter, as the supply of gas is now
measured, and this has been carried into effect in
places. There are, however, serious objections to this
method. One of the objections is that the very class of
persons whom it is desired to induce to use a plentiful

---

* Personal communication of Mr. R. K. MARTIN, Chief Engineer to the Water
Board.

† BUCK's Hygiene and Public Health, Vol. I, p. 214.

supply of water, would from motives of economy, use less than is necessary for cleanliness and health. A system of vigilant inspection of the water service in houses would probably serve to reduce this unnecessary waste to a considerable extent.

SOURCES OF DRINKING WATER.

All water, from whatever direct source odtained, comes originally, by precipitation, from the atmosphere. In many places the rain or snow water is the only source of supply. This is usually collected as it falls upon the roofs of buildings and conveyed by gutters and pipes to cisterns, where it is stored until needed.

In Venice, the rain falling upon the streets and court-yards is also collected in cisterns after filtering through sand. The cisterns used for the storage of water in New Orleans and other southern cities in the United States, where the temperature rarely falls below the freezing point, are generally constructed of wood and placed above ground. Farther north, where it is necessary to protect them against the action of frost, they are placed underground. These underground cisterns are usually built of brick. The water from cisterns above ground becomes very much heated, in summer, and necessitates the use of large quantities of ice to make it palatable. The water from the underground cisterns is pleasantly cool in summmer, and is also guarded against freezing in winter. There are, however, very serious objections to storing drinking water in underground cisterns. These reservoirs are usually placed within a few feet of privies and cess-pools, and as neither the retaining walls of the cisterns, nor those of the privies are water-tight, it often happens that the drinking water becomes strongly impregnated with the soluble portions of the excrement, or the products of its decomposition, which have drained into the cistern.

Personal observations in Memphis in 1879, as well as the careful chemical analyses made afterward by Dr. Chas. Smart, U. S. A.,[*] have convinced the author that the objections to all underground cisterns built of brick, stone, or cement, are insuperable from a sanitary point of view. Dr. Smart found over one-half of the underground cisterns examined by him in Memphis and other cities and towns to be leaky and presenting evidence of organic pollution. The water from 31 out of 80 cisterns analyzed, showed decided contamination by sewage. It would seem advisable to prohibit all underground cisterns for the storage of drinking water unless they are constructed of iron, which should be protected against oxidation by a thorough coating of coal tar. Where any other system of collection and storage is available, however, the underground cistern should be unreservedly condemned.

Rain water collected in the country, away from manufacturing districts, is usually quite pure and wholesome. Its taste is however flat and insipid, owing to absence of carbonic acid gas and mineral constituents. In cities rain water frequently contains such a large amount of organic matter and other impurities, which have been washed out of the air by the rain, that it may be unfit for drinking. On account of its softness, rain water is, however, very desirable for washing and other domestic purposes. If the statement made in the last chapter, concerning the presence of organisms in the atmosphere is remembered, it will be evident on a moment's thought that such organisms, when contained in rain water may be the source of disease. The putrefaction which so readily takes place in rain water upon standing a few days is caused by certain of the organisms carried down out of the lower strata of the air by the descending rain or snow.

---

[*] Report National Board of Health, 1880, p. 437-441.

Precipitation is an exceedingly unreliable source of water, and should never be depended upon when other sources of supply are available. Water famines are frequent wherever people are compelled to rely upon such an uncertain source of supply as rain or snow.

Rivers and smaller streams probably supply the larger number of cities and towns in this country with drinking water. When care is taken to prevent the pollution of the stream above the point whence the water is taken, this is usually of fair quality for domestic purposes. When the river can be tapped near its source, or before a large number of manufacturing establishments can empty their waste products into its current, or before it receives the sewage of a considerable number of inhabitants living on its banks, the water can generally be regarded as safe. It is very difficult, however, except in the less settled portions of the country to find these favorable conditions present.

Among the minor objections to the use of river water for domestic purposes are the liability of most streams to become turbid in times of freshet, and the discoloration of the water from dissolved coloring matters if the stream flows through a marshy or peaty region. These objections are, however, not serious, as filtration will readily remove the suspended matters. The coloring matter is probably harmless. The organic matter contained in the water of some streams, even when pollution by sewage and manufacturing refuse is absolutely excluded, may, however, be the cause of disease. Dr. Smart has shown* that the water from streams in Nebraska, Wyoming and Utah contained organic matter varying in amount from .16 to .28 parts per million.† He thinks the so-called 'mountain fever'

---

* American Journal Med. Sciences, January, 1878, p. 28 et seq.

† The source of this organic matter seems to be the melted snow which makes up a large portion of the streams.

of the Rocky Mountain region is a malarial fever caused
by the large amount of organic matter in the drinking
water.

The most serious objection to the use of river water
for domestic purposes, is the employment of streams as
carriers of refuse from manufacturing establishments,
or of the sewage of cities and towns.   In Great Britain
and some parts of the continent of Europe, owing to the
density of population and the variety and extent of
manufacturing industries, many of the streams are in
an extremely filthy condition.   In this country too,
especially in the more thickly settled manufacturing
districts of New England, the pollution of rivers has
increased to a degree to seriously jeopardize the health
of the people who are compelled to draw their water
supply from such streams.   Several years since a commission was appointed by the State Board of Health
of Massachusetts to inquire into the extent of the pollution of the streams in that State, and to devise means
for preventing. such pollution.   The commission extended its inquiries and observations over several years,
reporting the result to the State authorities at intervals.*   It was found that the water of the Blackstone
river, at Blackstone, where it crosses the State line and
enters Rhode Island, contained over ten per cent. of
sewage and refuse matters.†   Other streams in Massachusetts showed similar pollution.   That the presence
of such excessive amounts of refuse matters renders the
water unsuitable for domestic purposes, must appear
evident.   It is probable, however, that the most dangerous of the polluting matters are the excreta of human
beings, especially those of patients suffering from certain specific diseases, such as typhoid fever, or cholera.

---

* Reports State Board of Health of Massachusetts for 1873, 1874, 1876, 1877,
1878, 1879, 1880.

† Report State Board of Health of Massachusetts, 1876, p. 145.

Only a few years ago it was a generally accepted theory that running water, though polluted by sewage, 'purified itself' after flowing a distance of twelve miles, and the comforting and reassuring doctrine is still held by many. Recent observations point to the conclusion, however, that the self-purification of rivers is not entirely to be relied upon. A certain proportion of the sewage, it is true, undergoes oxidation in the presence of light and air, and minute organisms,* and so becomes changed into other, possibly innocuous compounds. But at present it is not known what proportion, or what kind of organic matter does undergo this change. Another portion of the impurities is deposited upon the bottom and sides of the stream, having been only held in suspension and not dissolved in the water. A portion probably forms chemical combinations with other suspended or dissolved matters, and is changed into compounds which may be volatile and pass off into the air, or form insoluble precipitates.

The remainder is rendered less perceptible, or imperceptible to chemical means by dilution. Every stream has sources of inflowing water—feeders—which increase its volume and thus dilute any foreign admixture.

In view of these facts, the theory of the self-purification of streams, as formerly held, can no longer be regarded as true. But it is unquestionably true that running water does regain its purity if the inflow of sewage and other refuse is not excessive. It cannot be stated with confidence, however, when a stream, once polluted, becomes fit for use again. Moreover, as it is not possible by any practicable chemical treatment, or filtration, to make a polluted water absolutely wholesome, it is safer not to use as a source of domestic sup-

* MUELLER and FALK: Desinfection, in EULENBURG's Realencyclopædie d. ges. Heilkunde, vol IV., p. 68.

ply, a stream which is known to have been seriously
contaminated by sewage matters or other impurities.

The water from fresh-water lakes and ponds is generally to be preferred to river water for domestic use.
It is less liable to become turbid from time to time, and,
except in the case of small ponds, inflow of sewage is
not likely to cause fouling of the water to any serious
extent. When the supply can be drawn from large
lakes, as is done in Chicago and other cities on the
great lakes of the United States, no purer or better
source can be desired. In these cases the point whence
the water is taken, should be far enough from shore to
avoid possibility of sewage contamination. When the
water-supply is taken from small ponds, all sewage and
waste products from houses and factories must be rigidly
excluded; otherwise, diseases, attributable to the polluted water, are likely to arise among those using the
same.

The water in small lakes and storage reservoirs
sometimes becomes offensive in taste and odor. The
water-supplies of several of the large Eastern cities
have within the past three or four years at times had a
peculiar odor and taste somewhat resembling cucumbers.
After considerable study, Prof. Ira Remsen, of Baltimore, found the cause of this odor and taste in a minute fresh-water sponge, the *spongilla fluviatilis*. A
still more offensive odor, tersely described as the 'pig-
pen odor' is given to the water by the decay of certain
species of nostoc and other algæ. It is not known that
either these vegetable, or animal organisms, if present,
render the water prejudicial to health.

Ponds are often used as sources of ice-supply. It
was formerly supposed that in the process of freezing,
solid matters in the water were not included in the
block of ice when congelation occurred. Recent observations have shown the falsity of this assumption. In

1875, an outbreak of acute intestinal disease at Rye Beach, New Hampshire, led to an inquiry by Dr. A. H. Nichols, which disclosed the fact that the ice used contained a large percentage of organic matter.* The use of ice from a different source was followed by an almost immediate disappearance of the disease. Upon further investigation, it was discovered that the impure ice had been gathered from a small, stagnant pond into which a small brook carried large quantities of sawdust from several saw-mills. The water of the pond was loaded with organic matter, and in summer the gases of decay arising from it were very offensive. Chemical examination showed that the ice from this pond contained nearly 6 parts of organic matter in 100,000, while in pure ice the organic matter amounted to only .3 parts in 100,000. A similar investigation into the character of the ice furnished to the residents of Newport, R. I., was made under the auspices of the Sanitary Protection Association of that city. The ice, which was cut from ponds in the immediate neighborhood of the city, was found to contain an excessive proportion of organic matter. Large quantities of sewage and other impurities were discharged into these ponds.†

A careful series of experiments recently made by Dr. C. P. Pengra, of Michigan, shows‡ that the purification of the water by freezing is in no sense absolute. In experimenting with bacteria, infusoria and other organisms, he found that from 9 to 11 per cent. remained in the ice, and retained their vitality, so that when thawed, they rapidly multiplied, and there was no apparent loss of numbers. In the ordinary process of freezing, the upper portion is the purest, but if snow

---

* Report Massachusetts State Board of Health, 1876, p. 467.

† The Dangers of Impure Ice. *The Sanitarian*, May, 1882.

‡ Private communication to the author. The memoir of Dr. PENGRA will be published in the Report of the Michigan State Board of Health for 1884.

or rain fall upon the ice, and freeze, this upper layer will be found much more impure than the lower. Rational conclusions from these experiments are, that ice should not be gathered from an impure source, and that an early harvest of the ice should be encouraged.

Springs and wells supply the water for most persons not aggregated in large communities, as cities and towns. Even in the latter, no inconsiderable quantity of the water used for drinking and domestic purposes is derived from wells. Spring water usually comes from a source at a considerable depth below the surface, that is to say, the water has percolated through thick strata of soil, before reappearing at the surface. In its passage through the soil it has lost most of its organic matter, and perhaps taken up mineral and gaseous constituents in larger quantities. It may be so strongly impregnated with the latter, as to vitiate it for ordinary use and to render it valuable as a medicine. Ordinarily, however, spring water is clear, cool and sparkling, with a refreshing taste and uniform temperature, and in all respects an agreeable and wholesome beverage.

The character of well-water, on the contrary, is often justly open to grave suspicion. Being derived from those strata of the soil which are most likely to be contaminated by the products of animal and vegetable decomposition, the wholesomeness of the water is inversely proportional to the degree of saturation of the soil with the products of decay. It has been found by experiment that when organic matter largely diluted with water is allowed to percolate through soil, it undergoes a gradual decomposition in the presence of certain minute organisms, nitrates and nitrites being formed at the expense of the ammonia and other organic combinations. If, however, the soil is saturated with organic matter in excess, and in a state of concentration, putrefaction takes place, and the conversion of the organic matter into nitrates and nitrites is retarded. Hence, the drainage of diluted sewage through a stratum of porous soil, not already saturated with putrefying matters, has no especially bad significance, even if the liquid should reach a well used as a source of drinking water. It is probable that by the time the liquid portion of the sewage reached the well it would have arrived at that point when it could truthfully be termed pure water. At the same time it must be remembered that the purifying power of the soil cannot be relied upon if the supply of sewage or other animal or vegetable impurity is too abundant.

Distillation is sometimes resorted to for the purpose of procuring drinking water, especially at sea. Vessels now generally carry a still for this purpose. The principal objection to distilled water is its insipidity, due to the absence of carbonic acid gas and mineral constituents, which give to good drinking water its savor. Distilled water may be aerated, by passing it in fine streams through holes in the bottom of a cask, elevated so as to allow the water to pass through a considerable stratum of air. Lead is sometimes taken up from the distilling apparatus, and may cause lead poisoning in those using the water.

Drinking water is sometimes procured by melting snow or ice. It is not probable that water derived from these sources is unwholesome, although there is strong popular prejudice against it. Ice and snow may however contain large amounts of impurities, as already referred to,* and be, for this reason unfit for use.

The following qualities are desirable in water for drinking and domestic purposes.

1. The water should be colorless, transparent, sufficiently aerated, of uniform temperature throughout the year, and without odor or decided taste.

---

* See page 37.

2. The mineral constituents (magnesium and lime salts) should not be present in greater proportion than three or four grains per gallon. More than this gives to water that quality known as 'hardness.'

3. There should be but little organic matter present, and no living or dead animal or vegetable organisms.

4. The water should be entirely free from ammonia and nitrous acid, and should contain but very small quantities of nitrates, chlorides and sulphates.

5. It should contain less than one-twentieth of a grain of lead per gallon. A larger proportion is likely to be followed by lead poisoning.

IMPURITIES IN WATER.

The transparency and the color of water are affected by the presence of suspended or dissolved, mineral or organic matters. If, after standing for a time, the water deposits a sediment, this is dependent upon insoluble matters. If the sediment turns black when heated in a porcelain capsule over an alcohol or gas flame, it contains organic matter. If the sediment or residue effervesces upon the addition of hydrochloric acid, the presence of carbonates is indicated. Water may be colored by metallic salts or by vegetable matter. It may also contain large quantities of mineral or organic matter, or even living organisms, without perceptibly diminishing its transparency. For example, the ova of tape-worms may exist in water in considerable numbers, and yet remain perfectly invisible, except under the microscope.

The presence of sulphur compounds, or of various vegetable and animal organisms (sponges, algæ, etc.*) may give to water an unpleasant odor and taste. In the oil regions of this country most of the drinking water is contaminated with petroleum, which is very

---

* See page 36.

disagreeable to one unaccustomed to it. It is not probable that the small quantities of the oil imbibed with
the water have any deleterious influences upon the
organism.

Many works on hygiene fix a limit to the amount
of solid matter allowable in drinking water. The International Congress of Hygiene, at Brussels, fixed the
limit at 50 parts in 100,000. It is impossible, however,
to say of any particular specimen of water, that its
content of solid matter, whether organic or mineral,
will be prejudicial to health, without trial. At the
same time it is prudent to reject all waters containing
a considerable proportion of solid organic matter, as
determined by the degree of blackening or heating the
sediment, or residue after evaporation.

The hardness of water is due to the presence of
earthy carbonates, or sulphates, or both. If the hardness, is due to carbonates, it is dissipated by heat, as
in boiling the water; the carbonic acid is driven off,
and the base (calcium or magnesium oxide) is precipitated upon the bottom and sides of the vessel. This
is termed the 'removable hardness.' The hardness
due to the presence of earthy sulphates is not removed
upon heating the water, and is termed the 'permanent
hardness.' The hardness depending upon both the carbonates and sulphates is called the 'total hardness.'

The proportion of the above mentioned earthy salts
present in a given specimen of water, is determined by
what is called the soap-test. This test depends upon
the property which lime and magnesia salts possess of
decomposing soap, (oleate and stearate of soda). The
quantity of a solution of soap of a definite composition
decomposed by a quantity of hard water, indicates the
amount of the salts present. In this country and England, this is generally expressed in degrees of Clark's
scale, which are equivalent to grains of carbonate of

lime per imperial gallon. Thus, if the chemist says
that a certain sample of water has a total hardness of
16°, he means that the earthy salts in the sample decom-
pose the same quantity of soap that would be decomposed by 16 grains of carbonate of lime per imperial
gallon. In Germany each degree of the scale used ex-
presses the soap decomposed by one part of calcium
oxide per 100,000. In the scale used in France each
degree corresponds to 1 part of carbonate of lime in
100,000. So much of the hardness of water as is due
to carbonates, can be dissipated by boiling, which
drives off the free carbonic acid.

Hard water is objectionable for domestic use as it
is wasteful of soap. In cooking certain vegetables,
such as peas and beans, the hulls are not thoroughly
softened. In making infusions of tea and coffee, larger
quantities of these materials are needed than where
soft water is used.

DISEASES DUE TO IMPURE DRINKING WATER.

Hard water is popularly believed to be the cause of
calculous diseases, and of goitre and cretinism, but no
reliable observations are on record showing that the
belief is founded upon fact. At the same time it is
undoubtedly true that calcareous waters produce
gastric and intestinal derangements in those unaccustomed to their use.

Large amounts of suspended mineral matter are
frequently present in river water, and may give rise to
derangements of the digestive organs. If there is
carbonate of lime present, the water can be easily
clarified by the addition of a small quantity of alum.
Sulphate of lime and a bulky precipitate of hydrate
of alumina are formed, which carry the suspended
matters to the bottom. About six grains of crystallized
alum are sufficient to clarify a gallon of water. This

amount of alum is too small to affect the taste of the water perceptibly. This method is frequently used to clarify and render fit for use the water of the Mississippi river, which is usually very muddy. Dr. Parkes quotes the following striking instance of the practical value of clarifying muddy water by means of alum.* In 1868, the right wing of the 92nd Regiment of Highlanders, going up the river Indus, suffered from diarrhœa from the use of the water, which was very muddy. The left wing of the same regiment used water from the same source, but precipitated the suspended matters with alum, and had no diarrhœa. The right wing then adopted the same plan with like success. Although the opinion is wide-spread that water containing much mineral matter, either in solution or in suspension, is deleterious to health, there is very little evidence absolutely reliable upon this point.

The presence of large quantities of organic matter in water, whether these matters be of animal or vegetable origin, must always be looked upon with suspicion. The observation was made by Hippocrates twenty-three centuries ago, that persons using the water from marshes, *i. e.*, water containing vegetable matter, suffer from enlarged spleens. Many physicians, both of ancient and modern times seem to have held this opinion, but the first positive observation in medical literature is the now classical one of the ship *Argo*, reported by Boudin.† In 1834, the transport *Argo*, in company with two other vessels, carried 800 soldiers from Bona in Algiers, to Marseilles. The troops were all in good health when they left Algiers. All three of the vessels arrived in Marseilles on the same day. In two of them there were 680 men, not one of whom was sick. Out of the remaining 120 men who were

---

* Manual of Hygiene. 6th Ed., N. Y., Vol. I, p. 31.

† Quoted in PARKES. op. cit. p. 48. NOWAK: Lehrbuch der Hygiene. p. 51 ; and in numerous other publications on Hygiene.

on the third vessel, the *Argo*, 13 died during the passage, and 98 of the 107 survivors suffered from paludal fevers of all forms.   None of the crew of the
Argo were sick however.   The two vessels exempt
from sickness, and the crew of the *Argo* had been
supplied with pure water, while the soldiers on the
latter vessel had been furnished with water from a
marsh.   This water was said to have a disagreeable
odor and taste.   The testimony of a large number of
East Indian physicians is also quoted by Parkes in
support of the view that malarial fevers are often
caused by impure drinking water.   The observations of
Dr. Charles Smart upon the production of the 'mountain fever' of the Western territories, have already
been referred to.   The author ventures to state it as his
opinion however, that the instances in which malarial
fevers are due to impure drinking water are very rare.

The causation of typhoid fever and cholera by
impure drinking water will be presently referred to.
Recently the opinion has been expressed by some that
yellow fever and diphtheria are also spread by polluted
drinking water, but no strong evidence has yet been
adduced in its support.

There can be very little doubt that diarrhœa and
dysentery are frequently caused by water which has
been contaminated with decaying organic matter.   The
evidence in favor of this amounts practically to demonstration.

It must not be forgotten that the ova of certain
animal parasites, as distoma hematobium, filaria san-
guinis hominis, and medinensis, anchylostoma duode-
nale, and possibly of round and tape worms, are taken
into the system with the drinking water.

Organic detritus of various kinds, sewage, decomposing animal and vegetable matter, refuse from
manufacturing establishments, may be a source of pol-

lution of water and render it unfit for drinking or other
domestic purposes. It is, however, not certain that
water thus rendered unclean is prejudicial to health;
in fact Dr. Emmerich, of Munich has recently put his
skepticism on this point to a practical test. For two
weeks he drank daily from one to two pints of very
filthy water; in fact, nothing less than sewage. The
water was both chemically and physically exceedingly
impure. Several of the experimenter's patients par-
took of the same water without any ill effect. He even
claims that a gastric catarrh from which he was suffering when the experiment was begun, improved during
its course.*

The results of Emmerich's experiments, and of
other well-known observations seem almost conclusive
that the products of animal and vegetable decomposition, taken into the body with the drinking water, can
not be looked upon as certainly harmful. Should,
however, water containing such impurities, or even
water apparently pure, contain the germs of one of the
specific diseases: cholera, typhoid fever, or perhaps
yellow, malarial or scarlet fevers, or diphtheria, it is
probable that such disease would be communicated to
the consumer of the water.

Many instances are on record where outbreaks of
typhoid fever have been clearly attributable to pollution of the drinking water by the germ of the disease
from a previous case.

One of the most remarkable of these outbreaks is
that recorded by Dr. Thorne.† About the end of
January, 1879, typhoid fever began suddenly in the
adjoining towns of Caterham and Red Hill. Within
six weeks three hundred and fifty-two cases occurred.

---

* WOLFFHUEGEL: Wasserversorgung, in PETTENKOFER u. ZIEMSSEN's Handbuch
der Hygiene. I. Abth., II. Hlfte., p. 97.

† Report of the Medical Officer to the Local Government Board for 1879. Quoted
in FODOR: Hygienische Untersuchungen, etc., II, Abth. p. 261.

All other sources of the disease were excluded, except the drinking water, to pollution of which it was traced with almost absolute certainty. Caterham contained 558 houses and Red Hill 1700. Of the former, 419 and of the latter 924 draw their drinking water from a common supply, having its source in a well several hundred feet deep. The insane asylum with 2000 inmates, and the military barracks in Caterham used water from a private well. There was no typhoid fever among the last two communities. During January one of the workmen engaged in some excavation near the public well was taken ill with diarrhœa and fever—probably typhoid, but was still able to continue his work. His dejections were often voided where they were certain to become mingled with the water of the common supply. This man's diarrhœa began on January 5th, and continued until the 20th of the month, during which time he remained at work. On the latter date, he was compelled to quit work and take to his bed. Exactly two weeks from the beginning of the man's sickness, on January 19th, the first case of typhoid occurred in Caterham, and then rapidly increased. The first case occurred, therefore, just fourteen days—the incubative period of typhoid—after the presumed infection of the the drinking water by the dejections of the sick laborer, who had come from Croydon, where typhoid fever was at the time prevalent. Within two weeks from the appearance of the first case, the epidemic had reached its height, and then rapidly declined, disappearing almost entirely in a month after the outbreak. It was shown by Dr. Thorne that nearly all the houses in which the disease appeared were supplied with water from the source above mentioned, while other houses in the immediate vicinity of the infected ones remained free from the disease.

In 1874 there was an outbreak of typhoid fever in the town of Over Darwen, in which nearly 10 per cent. of the inhabitants were attacked. Here the source of the disease was also traced to an infected water supply.

Dr. Buchanan has shown that an outbreak among the students of the University of Cambridge, was likewise attributable to an infected water supply.

In this country the reports of the Boards of Health of the various States teem with accounts of localized outbreaks of typhoid fever referred to infected or polluted drinking water. In most instances the evidence furnished by the observers is not conclusive. In many, however, especially of those found in the Massachusetts and Michigan reports the fact of the communication of the disease in this manner seems unquestionable. One of these is as follows: Out of forty families, all using water from a certain well, there occurred 23 cases of typhoid fever. Out of forty-seven families living in the same neighborhood, but using water from different sources, only two had typhoid fever.* Dr. C. F. Folsom has published a very suggestive account of a house epidemic,† where nine cases in a single house, who all drank water from a well which was proven to be infected from a privy, were attacked by this disease.

The numerous cases of typhoid fever, which have been attributed to the use of infected milk may be included in this category. It is probable that the milk became infected either through polluted water used for the purpose of cleansing the milk vessels, or in diluting the milk. Mr. Ernest Hart has recorded‡ fifty epidemics of typhoid fever, fifteen of scarlet fever, and seven of diphtheria, the cause of which he has attributed to infected milk.

---

* Transactions Mich. Med. Society, 1883, p. 401.

† Boston Med. and Surg. Journal, vol. CII., p. 227, 261.

‡ Transactions Int. Med. Congress, 1881, vol. IV. p. 391.

It is not probable, however, that typhoid fever is always, or in the majority of cases spread through the medium of polluted drinking water, but in many of the instances on record, the relations between cause and and effect,—impure water and typhoid fever, have been so clearly made out as to no longer permit any doubt upon the question.

As it is with typhoid fever, so also with cholera. In a later chapter the origin and propagation of typhoid fever and cholera will be discussed more fully. At the present time only the relations of the drinking water to the spread of these diseases can be considered. In the instance to be presently noted, the connexion between the infected water on the one hand and the outbreak of cholera on the other is so clearly shown as to be almost equivalent to a mathematical demonstration. The facts in the case, were brought to light after a patient inquiry by a commission whose report drawn up by Mr. John Marshall, has made the occurrence classical. In 1854, the people of a well-to-do, and otherwise healthy district in the Eastern part of London suffered severely from cholera. Upon inquiry the fact was elicited that a child had died of cholera at No. 40 Broad street, and that its excreta had been emptied into a cess-pool situated only three feet from the well of a public pump in that street, from which most of the neighboring people took their drinking water. It was further discovered that the bricks of the cess-pool wall were loose and permitted its contents to drain into the pump-well. (It should be noted that the communication between the cess-pool and well was direct; that there was immediate drainage, not percolation through the soil). In one day 140 to 150 people were attacked and it was found that nearly all the persons who had the malady during the first few days of the outbreak drank the water from the pump. When the pump was closed to public use by the authorities, the epidemic subsided. The most singular case connected with this outbreak was the following: In West End, Hampstead, several miles away from Broad street, there occurred a fatal case of cholera, in a woman, 59 years old. This woman formerly lived in Broad street, but had not been there for many months. A cart, however, went daily from Broad street to West End, carrying among other things, a large bottle of water from the pump referred to. The old lady preferred this water to all others, and secured a daily supply in the manner stated. A niece who was on a visit to the old lady drank of the same water. She returned to her home, in a high and healthy part of Islington, was likewise attacked by cholera and died. There were, at this time, no other cases of cholera at West End, or in the neighborhood of these last two persons attacked.

Most of the English medical officers in India hold strongly to the view that cholera is spread by polluted drinking water, and the evidence in its favor is very strong.

The evidence in favor of the influence of impure drinking water on the causation of other diseases than those mentioned is not sufficient to justify any conclusions at present.

The source of a water-supply may be pure, and yet pollution may occur before the water is used by the persons to whom it is distributed. Supply pipes may become defective and the water become contaminated with sewage or other deleterious substances. It is a current belief that no impurity can gain access to hydrant pipes between the reservoir, or source of supply, and the point of discharge of the water. Nevertheless, such contamination may occur very readily. The author and his colleague, Dr. J. W. Chambers, of Baltimore, proved this conclusively a few years ago, by establishing an undoubted connexion between a house-epidemic of typhoid fever and a defect in the hydrant supplying the family with water.* The hydrant was one of the class known as Clark's patent non-freezing hydrant. The mechanism of these hydrants is as follows: At the lower end of the vertical discharge pipe is a glazed earthenware plunger, which works through a ring of rubber packing into a vacuum chamber. At the bottom of the vacuum chamber is a valve regulating the entrance of the water from the conducting pipe. When the water is shut off, this valve is kept closed by a spiral spring.—When the crank of the hydrant is turned forward, that is, when the water is 'turned on,' the plunger is forced to the bottom of the vacuum chamber, presses on the spring, opens the valve and allows the water to discharge. When the crank is turned back, the plunger is raised, releases the spiral spring, which forces the valve into its bed, and shuts off the water. The partial vacuum produced by the raising of the plunger, draws the water which is in the vertical discharge pipe into the vacuum chamber, which is so far below the surface as to be unaffected by frost. In course of time, and with use, the rubber packing gets worn and permits gradual leakage into the vacuum chamber, of the dirty, stagnant water by which this part of the hydrant is always surrounded. Outbreaks of typhoid fever having a similar origin, in which the connexion between cause and effect was clearly shown have been reported by other physicians of the same city.†

Aside from the practical question of the causation of disease by polluted water, a more abstract and æsthetic idea is involved in consciously taking any impu-

---

* On Preventable Pollution of Hydrant Water and its Relation to the Spread of Typhoid Fever. Maryland Med. Journal, Vol. VII., p. 271.

† Local Causes of Insanitation in Baltimore, By JOHN MORRIS, M. D.. Report Md. State Board of Health, 1878.

rity into the system. The instincts of man as well as of most animals revolt at it. These inborn instincts which constitute the sanitary conscience, as Soyka says, demand purity of food and water, as they insist on cleanliness of the body, of clothing, and of the dwelling.

STORAGE AND PURIFICATION OF WATER.

Wherever a large supply of water is needed, unless drawn direct from a well or spring, or pumped directly from its source, arrangements for storage are necessary. Cisterns and large reservoirs are made use of for this purpose. River water, especially, requires a period of rest in a storage reservoir in order to allow deposition of the large amount of suspended matter in it. Prolonged storage also gives opportunity for the conversion of possibly deleterious organic compounds into simple, and perhaps harmless, compounds. Usually in an elaborate system of water-works, a series of reservoirs is built in which the water is stored successively, so that before its final distribution through the street mains, it has become quite clear and pure. Filtration on a large scale is also used in connexion with storage reservoirs, in order to secure greater purity of the water.

In the distribution of water, care should be taken that nothing deleterious is taken up by the water in its passage through the pipes. Lead poisoning is not infrequent from drinking water that has passed through a long reach of lead pipe, or which has been standing in a vessel lined with lead. Tanks and storage cisterns should therefore not be lined with lead, and the use of lead pipe in the supply-service should be avoided as much as possible. Fortunately most natural waters possess a considerable proportion of carbonic acid, which forms with the lead an almost insoluble carbonate of lead. This carbonate of lead is deposited on the inside of the pipes, and protects both the pipes against erosive action from other constituents of the water, and also prevents the contamination of the water by the lead. An excess of carbonic acid in the water renders this deposit soluble, and may cause serious poisoning. Any water which is shown by analysis to contain over one-twentieth of a grain of lead per gallon is dangerous and should be rejected.

Owing to the possibility of defilement of the water from improper construction of hydrants, all out-door hydrants should be discouraged as much as possible, and should be replaced by a simple tap-cock indoors. The pipes should also be laid deep enough under-ground, or otherwise protected against freezing in winter.

A number of methods, all more or less efficient, have been introduced to purify water, when it needs purification before being fit for use. These methods either comprise filtration, or seek to purify the water without the aid of this process. One of the methods of purification without filtration, consists in exposing the water to the air in small streams. This was proposed by Lind more than a century ago, and has since been frequently revived. The water is passed through a sieve, or a perforated tin or wooden plate, so as to cause it to fall for a distance through the air in finely divided currents. By this process sulphuretted hydrogen, offensive organic vapors, and possibly dissolved organic matters are removed. This process has been used in Russia on a large scale.

By boiling and agitation, carbonate of lime, sulphuretted hydrogen and organic matter are removed, or rendered innocuous. Vegetable germs are usually destroyed, although Tyndall has shown that some bacterial germs withstand a temperature higher than that of boiling water.

As has already been mentioned,* alum is one of the readiest and most efficient means of removing suspended matters from water.

Permanganate of potassium is sometimes used to purify water containing considerable organic matter. The permanganate rapidly oxidises the organic matter and is believed to render it harmless. There is no certainty, however, that the germs of specific diseases are destroyed by the action of this salt, in the proportion in which it could be used for the purposes of water purification.

A yellow tint is given to the water by the permanganate, which is due to finely divided peroxide of manganese. This does no harm, but is unpleasant.

Water, unfitted for use by organic matter, is sometimes rendered usable by infusing certain vegetable astringents in it. Thus, it is said that the Chinese drink water only in the form of tea, in certain parts of China, where the water contains large quantities of organic matter. The tannin of the tea-leaves precipitates the suspended matters and renders the water fit for use. Mixing the water with red wine, which is astringent, has the same effect.†

Filtration is an efficient means of removing suspended matters. Charcoal, sand, gravel and spongy iron are used as filtering material. The most economical filter is one made of fine, clean sand, above which layers of gravel of a gradually increasing size are placed. The coarser particles of suspended matter are arrested before the sand, which removes most of the coloring and organic matters, is reached.

TESTS FOR IMPURITIES IN WATER.

Accurate and reliable quantitative analyses of water can only be made by chemists of experience. Every

---

* See page 42.

† CHAMPOUILLON, quoted in Med. and Surg. Hist. of the War, pt. II., Med. Volume, p. 613.

intelligent person should however know how to determine the presence or absence of suspected impurities.
The following methods are simple and easily carried
out:

To determine the presence of *chlorine*, or chlorides,
(often indicating sewage contamination), acidulate
about half an ounce of the water to be tested with pure
nitric acid, and add a few drops of a solution of nitrate
of silver, (24 grains to one ounce of distilled water).
A white precipitate, gradually changing to gray, is
produced if chlorides are present. The degree of cloudiness produced will indicate approximately, the amount
of chlorides. 'One grain of chlorine per gallon gives
a haze ; four grains per gallon give a marked turbidity ;
ten grains, a considerable precipitate.'* If the chlorine is found by this test to exceed one grain per gallon, the source of the contamination should be searched
for. If drainage from a cess-pool is suspected, a quantity of salt water may be thrown into it, and the water
again tested after an interval of four hours to see
whether the chlorine has increased.

The presence of *nitric acid*, or nitrates, even in
very minute quantities, is shown by the following test:
A small quantity of the water is evaporated to dryness, and a few drops of a solution of carbolic acid in
four parts of concentrated sulphuric acid and two parts
of distilled water, added to the residue. If nitric acid is
present, a brownish red color results, which turns green
and then yellow upon the addition of ammonia.

Nitrous acid or nitrites will give a reaction with
iodide of potassium and starch. Twelve to twenty
ounces of water in a flask are acidulated with a few
drops of dilute sulphuric acid, and a little solution of
iodide of potassium added. About half a dram of fresh-
ly prepared starch is added and the mixture shaken.

* PARKES, l. c., p. 77.

If nitrous acid is present, the iodide is decomposed setting free the iodine, which combines with the starch, causing a blue color. The test is a very delicate one.

The following test will indicate the presence of *ammonia*. Make a solution of one part of bichloride of mercury in thirty parts of water, and another of one part of carbonate of potash in fifty parts of water. Five drops of each of these solutions are added to about 24 ounces of the water to be tested, in a tall glass. If ammonia is present, a white cloudiness will appear in the water. If, after an interval of several hours, no turbidity has developed, no ammonia is present.

To determine the presence of *sulphuric acid*, or sulphates, the water is acidulated with hydrochloric acid, and a solution of chloride of barium added, which gives a white precipitate of sulphate of barium if sulphuric acid is present.

Organic matter is indicated by its reaction with chloride of gold, or potassium permanganate. A solution of permanganate added to water containing organic matter becomes almost instantly decolorized. The addition of chloride of gold, and boiling the water produces first a pink color, gradually changing through violet to black. The water should be first slightly acidulated.

Neither of these tests for organic matter is absolutely reliable, as nitrous acid, sulphuretted hydrogen, and ferrous salts produce similar reactions.

SIGNIFICATION OF THE VARIOUS IMPURITIES INDICATED.
BY THE ABOVE TESTS.

The following summary gives, in a brief compass, the inferences that may be drawn from the result of the above tests :*

---

* PARKES' Hygiene, vol. I, p. 79.

'If chlorine be present in considerable quantity, it either comes from strata containing chloride of sodium or calcium, from impregnation of sea-water, or from admixture of liquid excreta of men and animals. In the first case the water is often alkaline from sodium carbonate; there is an absence, or nearly so, of oxidized organic matters, as indicated by nitric and nitrous acids and ammonia, and of organic matter; there is often much sulphuric acid. If it be from calcium chloride, there is a large precipitate with ammonium oxalate after boiling. If the chlorine be from impregnation with sea-water, it is often in very large quantity; there is much magnesia, and little evidence of oxidized products from organic matters. If from sewage, the chlorine is marked, and there is coincident evidence of nitric and nitrous acids and ammonia, and if the contamination be recent, of oxidisable organic matters.

'Ammonia is almost always present in very small quantity, but if it be in large enough amount to be detected without distillation it is suspicious. If nitrates, etc., be also present, it is likely to be from animal substances, excreta, etc. Nitrates and nitrites indicate previously existing organic matters, probably animal, but nitrates may also arise from vegetable matter, although this is probably less usual. If nitrites largely exist, it is generally supposed that the contamination is recent; the coincidence of easily oxidized organic matters, of ammonia, and of chlorine in some quantity, would be in favor of an animal origin. If a water gives the test of nitric acid, but not of nitrous acid, and very little ammonia, either potassium, sodium or calcium nitrate is present, derived from soil impregnated with animal substances at some anterior date. If nitrites are present at first, and after a few days disappear, this arises from continued oxidation into nitrates; if nitrates disappear, it seems probable this is caused by the action of bacteria, or other low forms of life. Sometimes in such a case nitrites may be formed from the nitrates. Lime in large quantity indicates calcium carbonate if boiling removes the lime, sulphate or chloride or nitrate if boiling has little effect. Testing for calcium carbonate is important in connexion with purification with alum.* Sulphuric acid in large quantity with little lime, indicate sulphate of sodium, and usually much chloride and carbonate of sodium are also present, and on evaporation the water is alkaline. Large evidence of nitric acid, with little evidence of organic matter, indicates old contamination; if the organic matter be large, and especially if there be nitrous acid as well as nitric present, the impregnation is recent.'

[The following works are recommended to those desiring fuller information upon the subjects embraced in the foregoing chapter:

Water Supply, by WM. RIPLEY NICHOLS, N. Y., 1884. A Guide to the Microscopic Examination of Drinking Water, by J. D. MACDONALD, R. N. F. R. S. Some of the Organic Impurities found in Drinking Water, by C. W. CHAMBERLAIN, M. D., Rep't Ct. State Board of Health, 1880, p. 260. Microscopical Examination of Potable Water, by W. J. LEWIS, M. D., Ibid., p· 216.]

---

* See page 42.

CHAPTER III.

FOOD.

IN order to preserve health and vigor, it is necessary for animal beings to consume at intervals a sufficient quantity of substances known as foods. Alimentary substances, or foods, may therefore be briefly defined as materials, which, taken into the body and assimilated, sustain the processes of life, promote growth, or prevent destruction of the organized constituents of the body.

QUALITY AND CHARACTER OF FOOD NECESSARY.

It has long been known, as the result of the empirical observation of feeding large bodies of people, that the various proximate principles must be combined in certain definite proportions in the food in order to preserve the normal degree of health and vigor of the body. Within a comparatively recent period, physiologists have made experiments upon animals and human beings, which have led to the same conclusions, and have enabled these proportions to be fixed with more or less exactness.

Considering man as an omnivorous animal, it may be laid down as an invariable rule, that the following four alimentary principles are necessary to his existence.* Neither of these principles can be dispensed with for a prolonged period without illness or death resulting.

1. *Water.*—This must be supplied in sufficient quantity to permit the interchange of tissue to be carried on in the body.

---

* LANDOIS, Physiologie, 2te Aufl. p. 448.

2. *Salts.*—Inorganic compounds of various kinds are necessary to the preservation and proper construction of the tissues. They are all found in sufficient quantities in the various alimentary substances consumed by man and the lower animals. A deficiency of inorganic constituents in the food is followed by disease.

3. *Proteids.*—Organic nitrogenous material, either animal or vegetable, is a necessary constituent of the food of man. Continued existence is impossible without a sufficient supply of nitrogenous substances.

4. *Fats, or Carbo-hydrates.*—The organic non-nitrogenous, or carbonaceous principles of food are also necessary to the continuance of health. They are supplied either by fats, or by carbo-hydrates, (sugar, starch, etc.,) which may be used as substitutes for each other. Voit has shown that seventeen parts by weight, of starch, is equivalent as carbonaceous or oxidisable food to ten parts of fat.

According to Moleschott, the standard diet for a man of average height and weight, doing moderate work, should be as follows: Proteids, 130 grams (4.59 oz.); fats, 84 grams (2.96 oz.); carbo-hydrates, 404 grams (14. 26 oz.). The proportion of nitrogenous to non-nitrogenous food-stuffs would therefore be 1:3¾, or speaking roughly, 1:4. When very hard work is required, as from a prize-fighter in training, the proteids are about doubled, the fats increased by about one-fourth, and the carbo-hydrates decreased by two-thirds.

In addition to maintaining a proper proportion between the various alimentary principles, it is necessary to vary the articles of food themselves, otherwise they are liable to prove nauseating. The necessity of variety in the food, in order to preserve the appetite, is familiar to everyone.

The table here given,* shows the relative proportions of nitrogenous to non-nitrogenous alimentary
principles in the various substances named:

	NITROGENOUS.	NON–NITROGENOUS.
Veal, - - - - - - -	10.	1.
Beef, - - - - - -	10.	17.
Lentils, - - - - - -	10.	21.
Beans, - - - - - -	10.	22.
Peas, - - - - - - -	10.	23.
Fat Mutton, - - - - -	10.	27.
Pork, - - - - - - -	10.	30.
Cow's Milk, - - - - - -	10.	30.
Human Milk, - - - - -	10.	37.
Wheat Flour, - - - - - -	10.	46.
Oat Meal, - - - - - -	10.	50.
Potatoes, - - - - - -	10.	86.
Rice, - - - - - - -	10.	123.
Buckwheat Flour, - - - -	10.	130.

By keeping these proportions in view it will be seen
at once that if a man wished to live on beef alone, he
would have to eat four and a quarter pounds per day,
in order to get a sufficient amount of non-nitrogenous
food. Of potatoes, in order to get enough nitrogenous
food he would have to eat daily eighteen pounds. No
human stomach could prove equal to the task of digest-
ing this excess of material. On the other hand, it is
to be noted how perfect the combination of the various
principles is in human milk. In cow's milk, which is
nearest in composition to human milk, the non-nitrogenous principles are deficient. Hence, the important
practical point that when ordering milk diet for a patient a small portion of carbonaceous food, (bread)
must be added, if the standard of health shall be
reached, or maintained.

Climate has probably very little influence upon the
amount of food required by the individual. The
actual quantity of food consumed varies little between
various races, or in different parts of the earth. It is
true, however, that a larger proportion of fat is required in cold climates. That fatty articles of food

---

* LANDOIS, op. cit., p. 449.

readily undergo oxidation and furnish a large amount of animal heat is proven both by observation and experiment.

The albuminoid proximate principles of the food, proteids, are represented by the nitrogenous constituents of organic tissues. These are the vitellin and albumen of eggs, albumin, fibrin, globulin, myosin, syntonin and other nitrogenized principles of flesh and blood; the casein of milk, the gluten, fibrin and legumin of cereal and leguminous seeds and plants, gelatin and chondrin.

Fat constitutes an integral component of animal tissue, and is found in abundance as a constituent of nerve tissue, marrow, and sub-cutaneous connective tissue. In food it is represented especially, in the fatty tissue of meat, the yelk of eggs, butter, etc.

The carbo-hydrates are represented especially by various products of the vegetable world, as sugar, starch, dextrin, etc.

Water and the various other inorganic proximate principles, chief among which are compounds of calcium, sodium and potassium, are usually found in sufficient proportion in the other alimentary substances.

The food should be taken in appropriate quantities and properly prepared, A larger quantity than necessary may overtax the digestive organs and thus yield less than the required amount of nutritive material to the body.

Physical exertion increases the consumption of albuminoid and fatty principles. Hence, as in the cases of the athlete or prize-fighter in training, larger quantities of these principles are required to keep the nutrition of the body at the standard of health. During mental work, however, less nitrogenous material is consumed than during physical labor.

In youth the processes of combustion (production of carbonic acid) go on with greater rapidity than after adult life is reached. For this reason young persons rarely get fat,. the fat-producing food being burnt up in the body by the greater metabolic activity of the young cell. Hence, fats and carbo-hydrates should form a larger relative proportion in the diet of the young, than in that of grown persons.

Low external temperature causes a greater and more rapid consumption of fat than high external temperature. During febrile conditions, however, the destruction of the stored-up fat in the body—the wasting away—is one of the most notable phenomena. Hence, the importance of supplying fat and fat-producing food in chronic febrile diseases.

'Der Mensch ist was er isst,' said Ludwig Feuerbach.* The pungency of the epigram is somewhat lost in the translation, which is literally: 'Man is what he eats.' The intimate relations of mental, moral and physical conditions of health to the quality and quantity of food deserve the earnest attention of the educated physician and sanitarian.

FOODS.

Foods or victuals are generally divided into foods proper, and so-called accessary aliment. The classification is not exact however, as the latter, which are commonly regarded as articles of luxury, may under certain circumstances become necessities, and hence should not be considered as forming a separate class.

Foods are either of animal or vegetable origin. Those derived from animal sources are milk, the flesh of animals, birds, reptiles and fish, and the eggs of the three last-named.

---

* Gottheit, Freiheit und Unsterblichkeit von Standpunkt der Anthropologie, p. 5.

The foods derived from the vegetable kingdom comprise the seeds of various plants (cereals, legumes) roots, herbs, ripe fruits, the fleshy envelopes of various seeds (which may properly be classed with the fruits), and various fungi.

There are also in common use a number of beverages, *e. g.*, water, alcoholic liquors, alkaloidal infusions (tea, coffee, cocoa), etc.

In addition, a number of substances or compounds are in common use as condiments. Their function is either to render victuals more palatable, or to promote digestion and assimilation. Vinegar, mustard, and common salt, are familiar examples.

FOODS OF ANIMAL ORIGIN.

Milk.—Human milk is, so far as known, the one perfect food for man found in nature. It contains, in proper proportion, representatives of all the different classes of proximate principles necessary to nutrition. One hundred parts contain about 3.5 parts of proteids (casein and albumin); 4 parts of fat (butter); and 5 parts of sugar. The reaction of human milk is slightly alkaline; that of fresh cow's milk is neutral.

In human milk there are 12.21 parts of solid matter to 87.79 of water, while in cow's milk the proportions are, solids 11, water 89 parts.

Of the solids in milk, cow's milk contains more proteids, while human milk is richer in fats and sugar. Hence, in using cow's milk as a substitute for human milk, the proteids are diluted by the addition of water, and the non-nitrogenous components increased by adding sugar, and under some circumstances, fat (cream).

On standing, the fatty constituent of milk, the cream, separates, and on account of its less specific gravity, rises to the surface, where it forms a layer of varying thickness.

After standing a longer interval, the milk under-goes certain physical and chemical changes. Lactic acid is formed at the expense of part of the sugar of milk (a sort of fermentation taking place), and acting upon the casein, produces coagulation. This is the so-called 'bonny-clabber.' When the fermentation con-tinues, especially under a slightly elevated tempera-ture, the solid portion becomes condensed (curd), and a sweetish-acid, amber colored liquid, the whey, sepa-rates. The curd, after further fermentation, under appropriate treatment, becomes converted into cheese. Whey is sometimes used alone, or mixed with wine, as an article of diet for the sick.

Butter is made from the cream by prolonged agita-tion in a churn. The fat-globules adhere to each other and form a soft, unctuous mass, of a yellowish color, solid at ordinary temperatures. After the butter is all removed in this way the balance of the cream remains in the churn as buttermilk. This is an article of con-siderable nutritive value, although its excess of acid renders it unsuitable as an article of diet in many cases.

The specific gravity of fresh milk should not be below 1030. It should, however, be borne in mind that the richest milk is not always that which has the highest specific gravity. In fact, a sample of rich milk, containing a large proportion of cream, may show, when tested with the lactometer, a lower specific gravity than a specimen of much poorer milk. Hence, the lactometer, although a useful instrument in guard-ing against excessive dilution of the milk with water, is not a very reliable guide in determining the quality of the milk.

Objections are often urged against the use of so-called 'skim-milk,' i. e., milk from which the cream has been removed. In some cities in this country, the police, or representatives of the sanitary authorities, seize and confiscate all skim-milk found in possession of dealers. There appears to be no rational basis for the opinion held by many that skim-milk is not a proper and useful article of food. Before the lactic acid fermention has taken place, it differs from fresh milk merely in the fatty and other matters removed in the cream. It may be used as an article of food with great advantage and entire safety. In certain diseased states, it is of exceptional value, as an article of diet. The sole objection of any weight to skim-milk is that it may be at times sold fraudulently as fresh milk. This is, however, a question of little sanitary interest, but one principally of ethics.

Milk is frequently adulterated by the addition of water, chalk and water, or other more deleterious substances. An excess of water gives the milk a bluish tinge, and reduces its specific gravity. The addition of water may become especially dangerous by introducing the virus of some of the acute infectious diseases. Thus, localized epidemics of typhoid and scarlet fevers, have in a number of instances been traced to mixing the milk with water containing the poison of these diseases.

It is a mooted question whether acute or chronic infectious diseases of the milk-giving animal may be communicated to persons using the milk of such animals. Facts at present known seem to negative the proposition. At the same time it would seem to be prudent to avoid the use of milk from diseased animals, if possible, or to destroy any organic virus the milk may contain by previously boiling the milk. After thorough boiling little fear need be entertained of communicating either acute or chronic infectious diseases through this medium.

The milk of cows fed upon the refuse of breweries and distilleries,—'swill milk,' is believed by many physicians to be unwholesome. If so, it is probably only by reason of the unfavorable hygienic conditions under which the animals are kept. If the stables are clean, dry and well ventilated, and the animals receive plenty of fresh air and exercise, swill-fed cows should produce as nutritious milk as when they are fed upon different food. Much of the present agitation against 'swill milk' is more prompted by political demagoguery than by scientific knowledge.

In order to prevent the rapid fermentation of milk various methods of preservation have been adopted. The addition of alkalies, or antiseptics retards the lactic acid fermentation, while the abstraction of a portion of the water and addition of sugar (condensed milk) preserves it for an indefinite time. The mere addition of water restores it to nearly its original condition.

Butter.—Butter is of especial value as food on account of the large amount of easily digestible fat which it contains. It is almost always used as accessary to other articles of food, to render them more palatable. When pure and fresh, it is one of the most delicious of creature comforts. It soon undergoes the butyric acid fermentation, however, becoming 'rancid,' as it is termed, when it is unfit for food.

The great demand for butter and its consequent high price, have led to its extensive sophistication. Butter is now very largely substituted by an artificial product termed oleo-margarine or butterine. This artificial butter is made from beef-suet by the following process: Fresh beef fat is melted at as low a temperature as possible, never higher than 126°-128° F. All membrane and tissue are then removed, and the resulting clear fat is put into presses, where the stearine is extracted. The liquid fat, free from tissue, and with nearly all its stearine removed, is known as 'oleo-margarine oil.' The next step in the process is 'churning.' The oil is allowed to run into churns containing milk and a small quantity of coloring material (annatto), where, by means of rapidly revolving paddles, it is churned for about an hour. When this part of the process is complete, the substance is drawn off from the bottom of the churn, into cracked ice. When cool, it is taken from the ice, mixed with a proper quantity of salt, and is then worked like butter and put into firkins for the market. It is also moulded into attractive prints, in imitation of dairy butter.* When the materials from which oleo-margarine is made are sweet and clean, and when the process of manufacture is properly conducted, the resulting product is an entirely harmless article, and probably differs very little in nutritive value from butter itself.

Cheese.—The value of cheese as a food depends upon the large amount of proteids and fat which it contains. The rich varieties of cheese, such as Fromage de Brie and Roquefort, contain on an average thirty per cent. of fat, and twenty-seven per cent. of proteid compounds. Parmesan contains only about eight per cent. of fat, and nearly thirty-three per cent. of proteids, while Edam and Chester cheese which may be considered as standing about midway between the above, contain twenty per cent. of fat, and nearly twenty-eight per cent. of proteids. From these figures it appears that cheese is one of the most nutritious aliments obtainable, but it cannot be eaten in large quantities at a time, as it is exceedingly liable to cause disturbances of the digestive organs. The constipating property of cheese is well-known to the public.

Cheese is not often adulterated. The only articles

---

* DR. W. K. NEWTON, Fifth Annual Report of the State Board of Health of New Jersey, 1881, p. 107,

used with success in its sophistication are lard and oleo-margarine, which are incorporated with the casein during the process of, manufacture. It sometimes undergoes chemical changes which render it intensely poisonous when eaten.

Meat.—The flesh of mammals, reptiles, birds, fish, and invertebrate animals is used as food by man. Falck* has classified the varieties of animals which furnish food to the inhabitants of Europe. There are forty-seven varieties of the mammalian class, one hundred and five of birds, seven of amphibia, one hundred and ten of fish, and fifty-eight of invertebrates.

Meat is the most important source of proteids in the food. In the more commonly used varieties of meat, the proteids and fats constitute from twenty-five to fifty per cent. of the entire bulk, the proportion depending largely upon the age of the animal and its bodily condition. The following table shows the influence of these two factors upon the relative proportions of the fats and proteids contained in the meat:

Parts in 100.†

	PROTEIDS.	FATS.
Moderately Fat Beef.	21.39	5.19
Lean Beef.	20.54	1.78
Veal.	10.88	7.41
Very Fat Mutton	14.80	36 39
Fat Pork.	14.54	37.34
Lean Pork.	19.91	6.81
Hare.	23.84	1.13
Lean Chicken.	19.72	1.42

The flesh of animals, which is neutral in reaction immediately after death, soon becomes acid in consequence of the formation of lactic acid. The acid, acting upon the sarcolemma and the muscular fibre, renders it softer and more easily permeable by fluids when cooking, and more susceptible to the action of the gastric juice when the meat is taken into the stomach.

---

* Das Fleisch ; Gemeinverstændliches Handbuch der Wissenschaftlichen und Praktischen Fleischkunde.

† Abridged from LOEBISCH ; article 'Fleisch' in Realencyclopædie d. ges. Heilkunde. Vol. 5, p. 340.

Certain kinds of meat,—mutton and venison for example, are often kept so long before being eaten, that a considerable degree of putrefaction has taken place when they are brought upon the table. The wisdom of this practice is questionable from a hygienic point of view.

Meat is sometimes eaten raw, but it is usually first cooked. The methods of cooking in general use are boiling, frying, roasting, broiling and baking. By either of these methods of cooking, when properly carried out, the nutritious properties of the meat are preserved and it is rendered digestible. The culinary art deserves the closest attention of students of hygiene.

A number of soluble preparations of meat, (beef extract, beef essence, beef juice) are found in the market and highly recommended as containing all the nutritious qualities of the meat from which they are prepared. These, and similar products of domestic preparation, (broths and teas) contain in reality very little nutritive material, but are of use almost solely as stimulants to the appetite and digestion. They have a place in the dietary of the sick, but their nutritive value is small.

Meat may be unfit for food from various causes. Thus the flesh of animals dying from certain diseases; splenic fever, pleuro-pneumonia, tuberculosis in its advanced stages, cow or sheep-pox should not be used as food when it can be avoided. Cases are on record proving the poisonous character of meat from animals which suffered, at the time of death, from some of the above mentioned diseases. The most important condition to be borne in mind is that certain parasites, (trichina spiralis, echinococcus, cysticercus) which frequently infest the flesh of animals, especially hogs, not infrequently give rise to serious or even fatal diseases in persons consuming such meat. Any meat containing these parasites, or suspected of containing them, should therefore not be used as food unless precautions be first taken to destroy the life of the parasite.

Of the parasites mentioned, the trichina spiralis is the most important in this connexion, as it frequently occurs in the flesh of hogs, rats, dogs, cats and other carnivorous animals. Rats are said to be infested with the parasite more frequently than any other animals. The trichinæ are found in two forms, one the mature form, inhabiting the intestinal canal. The immature form, or muscle trichinæ, are found in striped muscle, coiled into spirals and encysted in a fibrous capsule. They gain access to their host in the following manner : Flesh containing living trichinæ is taken into the stomach, where the muscular tissue and the fibrous envelope are dissolved, and the enclosed worms set free. These mature in the intestinal canal, where sexual reproduction takes place, and the young embryos pass through the intestinal walls and other tissues until they become imbedded in striated muscle. Localized epidemics of trichinosis have been reported in this country and Europe, and in nearly every instance the source of the disease has been traced to the ingestion of uncooked pork. Meat known to be trichinous should not be used unless in times of great scarcity. It may, however, be rendered innocuous by thorough cooking. A temperature of 140°-160° F., destroys the life of the parasite and renders the meat safe. On account of the frequent occurrence of trichinæ in pork, this meat should never be eaten unless thoroughly cooked. It has been ascertained that salted and smoked pork is not free from danger, as the parasites are not killed in the process of curing the meat. Hence ham and sausage should not be eaten raw, as the danger from these articles is almost equally as great as from fresh pork.

Certain animals can devour with impunity substances which are intensely poisonous to human beings.
The flesh of the animals may be impregnated with
these poisons, and cause serious and fatal illness in
persons partaking of it. In this way may perhaps be
explained the cases of poisoning sometimes following
the eating of partridges and other birds.

Crabs, oysters, lobsters and other shell-fish, when
eaten out of season sometimes give rise to severe gastro-
intestinal irritation. Prof. McSherry, of Baltimore,
says he has seen all the gastro-enteric and nervous
symptoms, called among the Spanish people of the
West Indies by the name of *siguatera*, to designate a
disease following the consumption of poisonous fishes,
induced by eating oysters unseasonably. What has
been said of oysters applies equally to lobsters and
crabs. These marine luxuries all change readily, and
if eaten out of season, or not perfectly fresh, are liable
to cause enteric disease, cholera morbus, or *siguatera*.*
Some persons are also affected by an idiosyncrasy, on
account of which they cannot indulge in shell-fish without the most serious discomfort, manifesting itself at
one time by a violent outbreak of urticaria, at another
by the gastro-intestinal affections above referred to.
These attacks may occur even when the food is entirely
fresh.

During the putrefactive processes in meat certain
poisonous compounds are formed in the tissues which
cause symptoms of poisoning in those who use such
meat as food. Numerous cases have been reported
where sausages, fish, canned meats and similar articles
have given rise to serious or even fatal illness. In
most of these instances, the meat has undergone some
degree of putrefactive decomposition when it was used.
It is believed that the poisoning in these cases is due

---

* Health and How to Promote it. New York, 1879, p, 143.

to the action of ptomaines, which are produced during the decomposition of meat in the absence of oxygen. These compounds have been shown by Selmi and others to be intensely poisonous. In some cases of poisoning by canned meats, it is not improbable that the symptoms were due to lead poisoning. Most of the cans used for preserving meats, vegetables and oysters, permit contact between the fluids within the can and the lead used for sealing them. Recently an improvement has been introduced in this respect, and cans are now made in which all contact between the solder and the contents of the cans is prevented.*

The prevention of disease from tainted meat is one of the most important problems of public hygiene. Food animals should be inspected by qualified inspectors before slaughtering, to exclude animals suffering from diseases that would vitiate the meat. When the meat is exposed for sale upon the dealer's stall, it should be again inspected, and all found unfit for use as food, confiscated and destroyed. Meat, in which the presence of trichinæ or other parasites is suspected should be examined microscopically.†

Eggs.—Although eggs contain a large amount of the proteid and fatty alimentary principles, their value as food has probably been greatly overrated. The savory taste and ready digestibility of eggs has, however ren-

---

* In a paper read before the New York Medico-Legal Society, Dr. JNO. G. JOHNSON, (*Sanitarian*, June, 1884), reported six cases of poisoning from eating canned tomatoes Dr. JOHNSON attributes the symptoms to the effect of muriate of zinc and muriate of tin, which he believes to have gained access to the can in the process of sealing. The muriate of zinc is used as a flux in soldering and the excess of acid dissolves the tin from the internal face of the cap. No chemical examinations were made, however, to demonstrate this opinion.

† The prevention of the diseases of animals by National and State authorities, is one of the most logical and thoroughgoing means of preventing disease from unwholesome meat. The American Public Health Association has for some years devoted considerable attention to the investigation of the diseases of animals, and means for their prevention. The Department of the Interior of the National Government has likewise made the diseases of cattle and hogs a subject of study and published some valuable reports thereon.

dered them a popular article of food. For obvious reasons, the eggs of the common barn-yard fowl are most frequently used, those of ducks and geese being far inferior in flavor to the first named, and being likewise less easily obtained.

The method of cooking eggs is generally supposed to have considerable influence upon their digestibility. According to Dr. Beaumont's experiments made on Alexis St. Martin, raw eggs are digested in one and a half to two hours, fresh roasted in two hours and fifteen minutes, soft boiled or poached in three hours, and hard boiled or fried in three and a half hours. These experiments are, however, of very little value as a basis for general conclusions. It is probable that a hard-boiled egg is quite as easily digested in the healthy stomach as a raw one, if care be taken to masticate it well and eat bread with it, so that it is introduced into the stomach in a finely divided state.

Eggs readily undergo putrefaction, when sulphuretted hydrogen is formed in them in large quantities. When this has taken place they are manifestly unfit to be used as food.

FOODS OF VEGETABLE ORIGIN.

Bread.—The various cereal grains, when ground into flour, are used in making bread. The flours of wheat, rye, barley, buckwheat and Indian corn are almost exclusively used in breadmaking. Of these, rye flour is richest in gluten—the nitrogenous principle—while corn meal contains most starch. The bran, or cortical portion of grain contains a larger percentage of proteid principles than the white internal portion, hence, flours made from the whole grain, (bran flour, Graham flour) if finely ground are more nutritious than the white flours. The latter are however more digestible.

Good bread should be light, porous and well baked. The lightness and porosity are due to carbonic acid gas, imprisoned in cavities of the dough during the process of bread making. By adding yeast to the dough a fermentation is caused in the latter, in consequence of which a portion of the starch is converted into sugar, and then into alcohol and carbonic acid. During the process of mixing the dough, the entire mass becomes permeated by the gas, which, on heating, expands and leaves the numerous large and small cavities throughout the loaf which indicate properly made bread.

Instead of yeast, some persons use leaven, which is simply a portion of fermenting dough, saved from a previous baking. A small quantity of this added to a mass of dough starts up the fermentation in a similar manner to that of yeast.

The production of carbonic acid by fermentation in the dough, goes on at the expense of part of the starch. It has been proposed, therefore, to supply the carbonic acid from without, thus saving the entire amount of the carbo-hydrates present in the flour. This is accomplished in two ways. First, by the use of some alkaline carbonate or bicarbonate (bicarbonate of sodium, carbonate of ammonium), the carbonic acid being set free on the application of heat, or secondly, by forcing the gas, previously prepared, into the dough by means of machinery.

Flour is not infrequently adulterated with chalk, pipe-clay and similar articles. These are easily detected by adding a mineral acid which produces effervescence when it comes in contact with the alkaline carbonate used as adulterant. Bakers often mix alum with inferior grades of flour. This imparts a greater degree of whiteness to the bread, and, in addition, enables it to retain a large proportion of water, thereby increasing the weight of the loaf.

Formerly diseased grain (ergotised rye) often caused outbreaks of disease when the flour made from the diseased grain was used in bread-making. At the present time such accidents rarely occur. In some parts of Italy it is said that an endemic disease,—pellagra,—is caused by the consumption of diseased Indian corn. The evidence in favor of this view is, however, not unquestioned.

Potatoes and rice are often used with satisfaction as substitutes for bread. They both contain a large proportion of carbo-hydrates, and when properly cooked, are very palatable and easily digestible articles of food. Indian corn(hominy) and oatmeal are likewise wholesome and nutritious foods of this class.

The leguminous seeds, (beans, peas, lentils) furnish a food containing a large percentage of proteids. According to the analyses of Kœnig* the average composition of the most frequently used legumes in the dried condition is as follows:

	BEANS.	PEAS.	LENTILS.	GROUND-NUTS.†
Water, per ct. - - -	13 6	14.3	12.5	6.5
Solids, " - - - -	86.4	85.7	87.5	93.5
Proteids, - - - -	23.1	22.6	24.8	28.3
Fats, - - - - -	2.3	1.7	1 9	46.4
Carbo-hydrates, - - -	53.6	53.2	54.7 }	15.7
Cellulose, - - - -	3.9	5.5	3.6 }	
Ash, - - - -	3.5	2.7	2.5	3.2

Beans, peas and lentils are often added to other articles of food with advantage. In recent years an important article of food for armies has been made of various legumes ground into flour and mixed with fat, dried and powdered meat, salt and spice. This constitutes the so-called 'Erbswurst,' which formed such an important part of the dietary of the German army in the Franco-German war of 1871. Bean and pea meal is also used sometimes as an addition to other flours in bread-making. The dried leguminous fruits

---

* Die Menschlichen Nahrungs-und Genussmittel, II., p. 288,

† The American pea-nut, the fruit or nut of *arachis hypogœa*.

cannot be used as regular articles of diet, however, as
they soon pall upon the taste, and produce indigestion,
nausea and other intestinal derangements.

Green Vegetables.—The plants usually classed together as 'vegetables,' the products of the market
garden or truck farm, comprise cabbages, turnips, pars-
nips, onions, beets, carrots, tomatoes, lettuce, green
peas and beans and similar articles. They all contain
a large proportion of water, a variable proportion of
sugar, and a small percentage of proteid principles.
Much of their palatability and digestibility depends
upon the method by which they are prepared for the
table. All garden vegetables should be used soon after
being gathered, as they rapidly undergo decomposition,
and are liable to produce derangements of the digestive organs if used under these conditions.

Fruits and Nuts generally contain large quantities
of sugar and fats. They form agreeable additions to
other articles of diet, but are insufficient to sustain
life. The use of fruits usually produces copious intestinal evacuations, and they are therefore especially to
be recommended to persons of sedentary occupations,
in whom torpidity of the bowels is so frequently present.

Condiments.—Various aromatic herbs and seeds
are used as additions to other articles of food to increase their sapidity, and to promote a larger flow of
saliva and gastric juice, and so assist digestion. Mustard, pepper, alspice and vinegar are the principal condiments. Within certain limits they are not injurious,
but the tendency in the use of all stimulants is to exceed the healthful limit. Condiments, as well as other
stimulants, should be used in moderation.

COOKING.

Much more attention than is generally given should
be paid by physicians to the culinary art. The man-

ner in which food is cooked has no little influence upon
its digestibility. There can be no question that the
extreme prevalence of functional indigestion in this
country is almost exclusively dependent upon bad
cooking.

The various methods of cooking are boiling, frying, roasting, broiling and baking. By either of these
methods food can be cooked so as to be palatable as
well as digestible; on the other hand the choicest
article can be utterly spoiled and rendered unfit to be
taken into the human stomach. It depends, therefore,
not so much upon the method of cooking, as upon the
knowledge and art of the cook.

Boiling.—Meats of all kinds are rendered tender
and digestible by boiling. In order to retain the flavor
of meat, the water should be boiling when the meat is
put into it. By the heat of the boiling water the albumen on the outside of the meat is coagulated and
the juices and flavor retained within. After a few
minutes the temperature of the water should be reduced
to 160-170° F., and maintained at that height until the
meat is tender. By this process a much more savory
piece of beef, mutton or fowl can be obtained, than
where the meat is put into cold water and this gradually heated. The latter method is, however, the
proper one to be followed when good soup or broth is
desired.

In boiling vegetables, as much care is necessary as
in boiling meat or fish. Potatoes and rice should be
steamed, rather than boiled.

The difficulty of obtaining a good cup of coffee,
especially in the Northern portion of the United States,
illustrates the prevailing ignorance upon one of the
simplest points in the art of cooking. Coffee should
never be served in the form of a decoction, that is to
say, it should never be boiled. Properly made it is an

infusion, like tea, which no one ever thinks of boiling.
The difference between an infusion (especially if made
by percolation) and a decoction of coffee can only be
appreciated by those who have enjoyed the one, and
endured the other.

Frying.—Frying, if properly done, is really nothing
less nor more than boiling in oil or fluid fat of some
kind.   Olive oil is preferable but is not essential; butter, beef-drippings, lard, or probably cotton seed oil
may be substituted for it without disadvantage.   The
principle of frying depends upon the fact that the temperature of oil can be raised to such a height as to produce instant coagulation of the surface of meat, fish or
other object immersed in it while hot; this film of coagulated albumen imprisons the juices and flavors of
the meat or fish, and prevents the fat entering and
soaking the fibres with grease.   Small fish or birds
properly fried are justly regarded as delicacies by con-
noisseurs, but the process of saturating these objects
with fat while gradually heating them produces a dish
that is anything rather than grateful to the palate, or
conducive to good digestion.

Roasting.—The fame of 'the roast beef of Old
England' has passed into song, but at the present day,
beef and other meats are rarely roasted, either in this
country or abroad.   As Sir Henry Thompson well ex-
presses it,* 'the joint, which formerly turned in a current of fresh air before a well-made fire, is now half
stifled in a close atmosphere of its own vapors, very
much to the destruction of the characteristic flavor of
a roast.'   It is probable that the old method of roasting before an open fire produced not only the most
savory, but likewise the most nutritious and digestible
meat.   It is much to be regretted that the process has
fallen so greatly into disuse.

---

* Food and Feeding, London, 1880, p. 45.

Broiling and Baking.—These methods of cooking are modifications of the process of roasting. Meats or fish carefully broiled or baked preserve their natural juices and flavors to a great extent, and retain their digestibility and nutritious properties. Of all methods of cooking these are probably best known and most satisfactorily applied in this country.

ALIMENTARY BEVERAGES.

The alimentary beverages may be divided into two classes. Those depending for their effects upon the alcohol they contain, and those whose active principles reside in certain alkaloids. They are used chiefly as digestive and nervous stimulants.

BEVERAGES CONTAINING ALCOHOL.

The physiological action of alcohol has been pretty fully worked out by Binz and his pupils, and by other experimenters. From these researches, it appears that the first effect of taking alcohol, sufficiently diluted, into the stomach, is to increase the flow of the saliva and gastric juice. This effect is probably reflex, and results from a stimulation of nerve terminations in the stomach. The alcohol is rapidly absorbed, and is carried in the blood, without undergoing chemical change, to the nervous centres, lungs and tissues generally. In the brain the alcohol probably enters into combination with the nervous tissue, modifying the normal activity of the various centres, either increasing the activity, if the alcohol is in small quantity, (stimulating effect) or diminishing it, if in larger quantity, (depressing effect) or entirely suspending the activity of the centres, if in sufficiently large quantity, (paralysing effect).

Alcohol stimulates the vaso-dilator nerves, causing dilatation of the smaller vessels; in consequence of

this the blood is largely sent to the periphery of the
body ; the blood pressure diminishes, and heat-radiation is increased. At the same time a portion of the
alcohol is used up in the lungs in the production of
animal heat, thus economising the expenditure of fats
and proteids, and acting as a true respiratory food.
Alcohol does not contribute nutritive material to the
body; it only permits that which is stored up to be
saved for other uses, by furnishing easily oxidisable
(combustible) material for carrying on the respiratory
process, and supplying animal heat.

During the use of alcohol the excretion of urea is
diminished. This shows that waste of tissue is retarded in the body.

Regarding the statement of some authorities that
alcohol does not undergo any change in the body, but is
excreted unchanged, Binz asserts* that alcohol appears
in the urine only when exceptionally large quantities
have been taken, and then in very small proportion.
It is not excreted by the lungs, the peculiar odor of
the breath being due not to the alcohol, but to the volatile aromatic ether, which is oxidised with greater
difficulty and so escapes unchanged.

While alcohol produces subjectively an agreeable
sensation of warmth in the stomach and on the surface
of the body, the bodily temperature is not raised.
The subjective sensation is due to the dilatation of the
blood vessels, and the sudden hyperemia of those parts.

During fevers and other exhausting diseases, alcohol is invaluable to prevent waste of tissue and sustain
the strength. It does not act merely as a stimulant to
the circulation and nervous system, but as above
pointed out, saves the more stable compounds by furnishing a readily oxidisable respiratory food.

---

* Realencyclopædia, d. ges. Heilk, Bd. I. p. 183.

When taken in small doses by healthy persons, alcohol diminishes the temperature by increasing heat-radiation. When large quantities are taken, the bodily temperature is reduced by diminishing heat production, as well as by increased radiation. This is shown in the condition known as dead-drunkenness, in which the temperature is sometimes depressed as much as 20° below the normal. Cases in which the temperature sank to 75°, 78.8° and 83° F. have been reported, with recovery in all the cases.

The constant use of alcohol produces in all the organs an excess of connective tissue, followed by fatty degeneration and the condition known as cirrhosis. The organs most frequently affected are the stomach, liver and kidneys. Serious pathological alterations also occur in the circulatory, respiratory and nervous systems.

Alcohol is not necessary to persons in good health. Probably most persons, regardless of their state of health, do better without it. Its habitual use in the form of strong liquors is to be unreservedly condemned. The lighter wines and malt liquors, if obtained pure, may be consumed in moderate quantities without ill effects. Even in these forms however, the use of alcohol should be discouraged, or perhaps prohibited in the young.

Neither in hot, nor in cold climates is alcohol necessary to the preservation of health, and its moderate use even produces more injury than benefit. The Polar voyager and the East India merchant are alike better off without alcohol than with it.

It has long been a prevalent belief that the use of alcohol enables persons to withstand fatigue better than where no alcohol is used. A large amount of concurrent testimony absolutely negatives this belief.*

---

* See PARKES' Hygiene, 6th Ed., Vol. 1, p. 315-327.

The predisposition to many diseases is greatly increased by the habitual use of alcohol. Sunstroke, the acute infectious diseases, and many local organic affections attack by preference, the intemperate. A recent collective investigation by the British Medical Association, brought out the fact that croupous pneumonia, is vastly more fatal among the intemperate than among those who abstained from the use of alcoholic liquors.

Alcohol as a beverage, is consumed in the various forms of spirits, wines, and fermented liquors. The varieties of spirits most frequently used are brandy, whiskey, rum and gin. They are all procured by distillation.

Brandy is distilled from fermented grape juice and has a characteristic aromatic flavor. When pure and mellowed with age it is the most grateful to the palate of all distilled spirits.

Whiskey is distilled from barley, rye, oats, corn or potatoes. Each of these has a peculiar flavor, depending upon the particular volatile ether formed during the distillation. Rye, barley and corn whiskeys are almost exclusively used in this country.

Rum is distilled from molasses, and is a favorite ingredient in hot punches. It is often used with milk, eggs and sugar, in the preparation of egg-nogg, a highly nutritious, stimulating drink, which is often prescribed with great benefit in acute and chronic wasting diseases.

Gin is an ardent distilled spirit flavored with oil of juniper. It has a widely-spread popular reputation as a cure for kidney diseases, but is probably oftener responsible for the production of these diseases than for their cure.

All of the above mentioned liquors contain from forty to sixty per cent. of alcohol, and should always be diluted before being taken into the stomach, in order to prevent the local irritant effects of the alcohol upon the gastric mucous membrane.

Wine is the product of the alcoholic fermentation of the saccharine constituents of fruits. Wine is usually derived from the grape, though other fruits may also furnish it. The stronger wines, (sherry, port, madeira) contain from sixteen to twenty-five per cent. of alcohol. The lighter wines (hock, red and white Bordeaux and Burgundy wines, champagnes) contain from six to fifteen per cent. of alcohol. Some also contain considerable free carbonic acid, (sparkling wines) of which the champagnes are types. The red and white Bordeaux and Rhine wines are probably the least objectionable of these beverages for habitual use. They contain sufficient alcohol to be lightly stimulant, have a pleasant acid flavor, and are least likely to produce the bad effects which usually follow in the wake of the habitual use of the stronger wines, or ardent spirits.

Preference should be given to the wines of domestic manufacture, on account of the great probability of adulteration of the favorite brands of foreign wines.

Cider is the fermented juice of apples. It frequently produces unpleasant gastric and intestinal disturbances when drunk, on account of the large quantity of malic acid contained in it. Although it is usually ranked as a 'temperance drink' it is quite capable of causing intoxication when consumed in large quantities.

Beer is the fermented extract of barley, mixed with a decoction of hops and boiled. It should be prepared only of malt, hops, yeast and water, and should contain from three to four per cent. of alcohol, five to six per cent. of extract of malt and hops, two to four per cent. of lactic and acetic acids, and from one-fourth to one-half per cent. of carbonic acid. This ideal is however rarely attained in the article sold by the liquor dealer. Numerous adulterations are practised on the unsuspecting consumer. The hops are frequently substituted by aloes, calamus and ginger, or by the more deleterious picric acid, or picrotoxin. The rich brown color, sweetness, body and creamy foam are produced by caramel and glycerine. The more expensive barley malt is substituted by starch and rice, or grape sugar and molasses.

Ale, porter and brown-stout are merely varieties of beer—some containing more sugar, others more extractive matter.

Beer and its correlatives have considerable dietetic value, owing not merely to the alcohol they contain, but largely to the sugar and acids entering into their composition. When used to excess, they often cause a considerable accumulation of fat.

Koumyss is the national beverage of the nomadic tribes of Tartary. It consists of the milk of mares which has undergone partial fermentation. Recently it has been introduced into Europe and this country, where it is made of cow's milk. It is a palatable, nutritious stimulant, and is often useful as a dietetic article in disease.

THE ALKALOIDAL BEVERAGES.

The virtues of the alkaloidal beverages depend upon certain alkaloids which differ very little in their chemical composition or physiological effects, and upon certain volatile aromatic constituents of the various articles used. The principal articles employed in the preparation of these beverages are coffee, tea, chocolate, maté and coca. It is estimated that 500,000,000 people drink coffee, 100,000,000 tea, 50,000,000 chocolate, 15,000,000 maté or Paraguay tea, and 10,000,000 coca.

All of these are active nervous stimulants and retarders of tissue-waste. They are all liable to produce serious functional disturbances of the nervous, digestive and circulatory systems if used to excess. Anemia, digestive derangements, constipation, pale, sallow complexion, loss of appetite, disturbed sleep, nervous headaches and neuralgias are the most marked of these effects.

On the other hand, when taken in moderate quantity, the alkaloidal beverages enable the consumer to withstand cold, fatigue and hunger; promptly remove the sensation of hunger, and diffuse a glow of exhilaration throughout the body.

Coffee.—Coffee is the ripe fruit (seed) of the *Caffea Arabica*, a native of Arabia and Eastern Africa, but now cultivated in other tropical regions of the world. The fruit consists of two flat-convex beans, the flat surfaces of which are apposed to each other. These are enclosed in a fibrous envelope which is sometimes used as a cheap substitute for the coffee-bean.

The beverage, coffee, is an infusion of the roasted and ground bean in hot water. Its virtues depend upon the alkaloid, caffein, and an aromatic oil. The latter, being volatile, is driven off by long-continued heat. Hence boiled coffee lacks the grateful aroma of that which is made by simply infusing the ground bean in hot water.

The great demand for coffee, and its comparatively high price have caused it to be extensively adulterated and substituted by other natural and artificial products. Artificial coffee beans have been made of clay, dough or extract of chicory, colored to imitate the natural bean. The fraud is easily detected by placing the beans in water, when the artificial product soon falls to pieces, while the natural beans undergo no change of shape or consistence.

Ground coffee as found in the stores is usually adulterated. The materials used for sophistication are, the grounds of coffee previously used, the roasted root of chicory, acorns, rye or barley, carrots, sun flower seeds, caramel and a number of articles of similar value, generally harmless.

Tea.—The plants which furnish the tea-leaves are natives of China, Indo-China and Japan. The tea-leaves contain a crystalline alkaloid, thein, identical in composition and properties with caffein. The various sorts of tea found in the market (green and black teas, etc.) differ only in the relative proportion of tannin and thein contained in each. The aromatic principle also varies somewhat in the different sorts.

Tea is adulterated to quite as great an extent as coffee, the leaves of various plants bearing more or less resemblance to tea leaves being added to the latter. Much of the tea found in the market is colored artificially with Prussian blue and iron oxide. These additions are harmless, as they are not soluble in water.

Chocolate.—Cocoa, from which chocolate is derived is widely different in composition from tea and coffee. In addition to its active principle, theobromin, which is identical with caffein and thein, it contains nearly 50 per cent. of fat, which renders it an article of high nutritive value.

Maté, or Paraguay tea, guarana, and coca are used to a considerable extent in some parts of South America, as substitutes for coffee and tea. Their composition is not well known, but their effects are believed to depend upon alkaloidal principles similar to caffein and thein.

TOBACCO.

Closely connected with the subjects treated in this chapter are the effects of the constant use of tobacco upon the human system. The depressing effects of tobacco, due principally to the nicotin upon the nervous and digestive systems, have long been recognized. Recently however, it has been found that very serious symptoms are produced upon the sense of vision by the constant or excessive use of tobacco. A special form of amaurosis, termed tobacco amaurosis, has been frequently noticed since attention was first called to it by Mackenzie.

[The following additional works are recommended to the student:

THOS. K. CHAMBERS, on Diet in Health and Disease. EDWARD SMITH, on Foods. FÖRSTER, Ernährung, in PETTENKOFER u. ZIEMSSEN's Handbuch der Hygiene.]

CHAPTER IV.

SOIL.

HIPPOCRATES treated at length, in one of his works, of the sanitary influences of the soil. Others of the older writers, especially Herodotus and Galen called attention to the same subject, and Vitruvius, the celebrated Roman architect, who flourished about the beginning of the Christian era, taught that a point of first importance in building a dwelling was to select a site upon a healthy soil.

From this time until the beginning of the eighteenth century, very little of value is found in medical literature bearing upon this subject. In 1717, however, Lancisi published his great work on the causes of malarial fevers in which he laid the foundation for the modern theory of malaria, and pointed out the relations existing between marshes and low-lying lands and those diseases, by common consent, called malarial. Other authors of the eighteenth, and the early part of the nineteenth century, refer to the connexion between the soil and disease, but exact investigations have only been made within the last thirty years.

When it is considered that the air that human beings breathe, and much of the water they drink, are influenced in their composition by the matters in the soil, the great importance of possessing a thorough knowledge of the physical and chemical conditions of the soil becomes evident to every one.

PHYSICAL AND CHEMICAL CHARACTERS OF THE SOIL.

In the hygienic, as in the geological sense, rock, sand, clay and gravel are included in the consideration of soils.

The soil, as it is presented to us at the surface of the earth, is the result of long ages of disintegration of the primitive rocks by the action of the elements, of the decomposition of organic remains, and possibly, of accretions of cosmical dust. The principal factor, however, is the action of water upon rock, in leveling the projections of the earth's surface produced by volcanic action.

Soils vary considerably in physical and chemical constitution. A soil may, for example, consist exclusively of sand, of clay, or of disintegrated calcareous matter. Other soils may consist of a mixture of two or more of these, together with vegetable matter undergoing slow oxidation. In forests, a layer of this slowly decomposing vegetable matter of varying thickness is found, covering the earthy substratum. This organic layer is called *humus*, and when turned under by plough or spade, and mixed with the sand or clay base, it constitutes the ordinary agricultural soil.

THE ATMOSPHERE OF THE SOIL, OR GROUND AIR.

The interstices of the soil are occupied by air or water, or by both together. The soil's atmosphere is continuous with, and resembles in physical and chemical properties that which envelopes the earth. Its proportion to the mass of the soil depends upon the degree of porosity of the soil and upon the amount of moisture present. In a very porous soil, such as, for example, a coarse sand, gravelly loam or coarse-grained sandstone, the amount of air is much greater than in a clayey soil, granite, or marble. So, likewise, when the soil contains a large proportion of water the air is to this extent excluded. The porosity of various soils, as evidenced by the amount of air contained in them, is much greater than would, at first thought, be supposed. Thus it has been found that porous sandstone may contain as much as one-third of its bulk of air, while the proportion of air contained in sand, gravel or loose soil may amount to from thirty to fifty per cent.

The ground-air is simply the atmospheric air which has penetrated into the interstices of the soil and taken part in the various chemical decompositions going on there. In consequence of these chemical changes the relative proportions of the oxygen and carbonic acid in the air are changed—oxygen disappearing and giving place to carbonic acid. It is well-known that during the decay of vegetable matter in the air, carbonic acid is formed; one constituent of this compound, the carbon being derived from the vegetable matter, while the oxygen is taken from the air. Hence, if this action takes place where there is not a very free circulation of air, as in the soil, the air there present soon loses its normal proportion of oxygen, which enters into combination with the carbon of the vegetable matter to form carbonic acid.

Thirty years ago, MM. Boussingault and Lévy, two distinguished French chemists, examined the air contained in ordinary agricultural soil, and found that the oxygen was diminished to about one-half of the proportion normally present in atmospheric air, while the carbonic acid was enormously increased. The exact results obtained by Boussingault and Lévy were as follows:

In one hundred volumes of ground air there were 10.35 volumes of oxygen, 79.91 volumes of nitrogen, 9.74 volumes of carbonic acid. In atmospheric air, on the other hand, there are in one hundred volumes 20.9 volumes of oxygen, 79.1 volumes of nitrogen, 0.04 volumes, or about one twenty-fifth of one per cent., of carbonic acid.

In spite of the striking results obtained by these two chemists, very little attention was paid to them by sanitarians, as very few seemed to have any clear notion of the relations existing between the motions of the air above ground and that under ground.

In 1871, however, Prof. Von Pettenkofer, of Munich, published the results of his own examinations into the constitution and physical conditions of the ground air, and the relations of the latter to the propagation of epidemic diseases. These researches, which created a wide-spread interest in the subject, were extended by other observers in all parts of the world. These observers, prominent among whom were Professors Fleck and Fodor, in Germany; Drs. Lewis and Cunningham, in India; Prof. Wm. Ripley Nichols, in Boston, and Surgeons J. H. Kidder and S. H. Griffith of the U. S. Navy, in Washington, demonstrated that the increase of carbonic acid in the ground air is due to increased vegetable decomposition and to lessened permeability of the soil. A permeable, that is to say, a sandy or gravelly soil, is likely to contain less carbonic acid in its atmosphere than a dense, less permeable clay, although the amount of decomposition going on, and the production of carbonic acid in the former may considerably exceed the latter. In the loose sandy soil, the circulation of the air is less obstructed, and the carbonic acid may easily escape and be diffused in the superincumbent air, while the close-pored clay imprisons the carbonic acid and prevents or retards its escape into the air above.

The disappearance of oxygen from the ground atmosphere is coincident with the production of an equivalent amount of carbonic acid. It appears from this that in the soil an oxidation of carbonaceous substances takes place, the product of which is the excess of carbonic acid in the ground air.

Prof. Nichols has found the proportion of carbonic acid in the air taken from a depth of ten feet below the surface in the 'made-land' of Boston, amount to 21.21 per thousand, the observation being made in August. In December, at a depth of six feet, the proportion was 3.23 per thousand. Fodor, in Buda-pesth, found the proportion of carbonic acid to be 107.5 per thousand (over 10 per cent.), the air being taken from a depth of thirteen feet.

Movements of the ground-atmosphere are principally due to differences of pressure and temperature in the air above ground. Owing to such differences the air from the soil frequently permeates houses, entering from cellars or basements. In winter, when the air of houses is very much more heated, (and consequently less dense) than the air out of doors, the difference of pressure, thus caused, draws the ground-air up through the house, while the cold external atmosphere penetrates the soil and occupies the place of the displaced ground-air.* A similar effect occurs in consequence of heavy rains. The water fills up the interstices of the soil near the surface and forces the ground-air out at points where the pores remain open. These places are the dry ground under buildings, where the air escapes and passes through floors and ceilings into the house above. Heavy rains may thus be the cause of pollution of the air in houses. The greater the porosity of the soil, the more likely is this to happen. This pollution of the house-air may be prevented by having impervious floors and walls to cellars and basements, or by interposing a layer of charcoal between the ground and the floor of the house. The latter does not prevent the passage of the ground-

---

* It is, of course, not strictly correct to say that the air *is drawn up* through the house by the diminution of pressure; it being rather *forced* out of the soil by the colder and denser outside air; but the phrase is sufficiently exact and will be readily understood.

air, but the charcoal layer absorbs the noxious matters, —filters the ground air, as it were.

In the spring and early summer the ground being colder than the air above it, and the ground-air consequently heavier and denser, the latter is not easily displaced. It is perhaps due to this fact that those infectious diseases which are probably dependent upon the movements of the ground-air are less prevalent in the spring and early summer than in the latter part of summer, autumn and early winter. In the autumn the ground-air being warmer than the air above ground is easily displaced by the latter and forced out into the streets and houses to be inspired by men and animals. The same conditions may explain the greater likelihood of infection at night, which is proven for such diseases as malarial and yellow fevers. The colder outside air penetrates the interstices of the soil and forces out the impure ground-air.

The researches of Fodor have demonstrated that the proportion of carbonic acid in the ground-air may be taken as an approximative measure of the impurity of the soil whence the air is taken. The influence of the permeability of the soil, as before pointed out, must however not be overlooked in estimating the significance of the carbonic acid. Fodor has shown that the proportion of carbonic acid in the ground-air, and consequently the amount of organic decomposition, is greatest in July and least in March. That the carbonic acid is derived from the decomposition of vegetable matter, has been proven by Pettenkofer. This observer examined specimens of air brought from the Lybian desert, and found that the proportion of carbonic acid in the ground-air was exactly the same as in the air collected above ground. There being no vegetable growth in the desert there can, of course, be no vegetable decomposition going on in the soil.

The excess of carbonic acid in the ground-air is an indication of the deficiency of oxygen as has been shown. The air at a depth of thirteen feet below the surface was found to contain only from seven to ten per cent. of oxygen—one-half to one third of the normal proportion. Many basements occupied by people as living rooms extend from five to ten feet underground, and hence are liable to be supplied with an atmosphere approaching in impurity that just mentioned. It requires no very vivid imagination to appreciate the dangers to health that lurk in such habitations.

THE WATER OF THE SOIL, OR GROUND-WATER.

At a variable depth below the surface of the ground, a stratum of earth or rock is found through which water passes with difficulty, if at all. Above this, there is a stratum of water which moves from a higher to a lower level, and which varies in depth at different times according to the amount of precipitation (rain or snow-fall) and according to the level of the nearest body of water toward which it flows. This stratum of water is termed the *ground-water*, and has within the last few years assumed considerable importance from its apparently close relations to the spread of certain of the infectious diseases. The direction of horizontal flow of the ground-water is always toward the drainage-area of the district. Thus, it is usually toward lakes, rivers or the sea. Rains, or a rise in the river cause a rise in the ground-water, while long continued dry weather, or a low stage of the river which drains off the ground-water causes a fall in the latter. On the sea-coast the ground-water oscillations probably correspond with the tides. The writer is not aware of any observations made to determine this point, with the exception of a single instance mentioned by Dr. De Chaumont. In Munich, where the ground-water flows

toward the river Isar, which divides the city, it has
been found that the annual range or oscillation (the
difference between the highest and lowest level during
the year) is ten feet, while the horizontal movement
amounts to fifteen feet per day. In Buda-pesth the
annual range was found by Fodor to be less than three
feet, while in some portions of India it amounts to
more than forty feet. As it is from the ground-water
that the greater portion of the supply of drinking water
in the country and in villages and small towns is drawn,
it becomes at once manifest how important it is to prevent, as far as possible, pollution of this source. Cess-
pools and manure-heaps and pits, of necessity, contam-
inate the soil, and also ground-water for a distance below and around them, and such water is clearly unfit
for drinking and other domestic purposes. Hence, the
reason why wells should not be placed too near privies
and manure-heaps or pits, becomes apparent.

Between the level of the ground-water, or that
portion of the soil where its pores are entirely occupied
by the water—where, in other words, the ground is
saturated—and the surface, is a stratum of earth more
or less *moist;* that is to say, the interstices of the soil
are partly filled with water and partly with air. It is
in this stratum that the processes of organic decay or
putrefaction are going on, in consequence of which the
pollution of the ground-air occurs. The oxidation of
non-nitrogenous matter in the soil results in the formation of carbonic acid. On the other hand, nitrogenized
compounds are oxidised into nitric acid and nitrates.
When, however, putrefaction occurs, nitrous acid, or
nitrites and ammonia are formed, the oxidation not
proceeding far enough to result in nitric acid.

Recent observations seem to show that these
processes of decomposition are initiated and kept up
by minute organisms termed *bacteria,* just as fermenta-

tion in liquids containing sugar can only take place in
the presence of the yeast plant. It has been found
that when non-putrefactive decomposition goes on, there
are always present multitudes of one variety of these
minute organisms; while if putrefactive decomposition
is going on, a number of other varieties of these
organisms are present. Just as, when a fermenting
liquid becomes putrid, the yeast plant disappears and
its place is taken by the ordinary bacteria of putrefaction, so in the soil, if the access of oxygen which is
necessary to the life of the bacteria of decay, is
prevented, these organisms die and are succeeded by
the organisms of putrefaction. It has been found that
in a soil saturated with water the bacteria of decay
cannot live, while those of putrefaction may flourish,
because these latter organisms can sustain life, and
develope in the absence of oxygen. Prof. Fodor's
researches indicate that the organism of non-putrefactive decomposition or decay is that which is termed by
Cohn *bacterium lineola;* and that the *bacterium termo*
is the principal organism of putrefaction.

DISEASES SPREAD BY SOIL IMPURITIES.

Given now an area of soil, say the ground upon
which a house or city is built, with a moist stratum in
which the processes of decay are active, and imagine a
rise in the ground-water. The ground-air, charged
with carbonic acid and other products of decomposition,
is forced out of the pores of the soil by the rising
ground-water, and escapes into the external air, or
through cellars and basements into houses and may
there produce disease. But the saturation of the soil
with water prevents the further development of the
bacteria of decay, and this is checked, or, putrefaction
may take place. If now, the ground-water sinks to its
former level or below, the processes of decay again

become very active in the moist stratum, and large quantities of carbonic acid ·and other inorganic compounds are produced. If the germs of infectious or contagious diseases have been introduced into the soil, they also multiply and may escape with the movements of the ground-air into the external atmosphere and there produce their infective action. This, it is held by Pettenkofer and his followers is what actually occurs in cholera and typhoid fever. Prof. De Chaumont has laid down the rule that a soil with a persistently low stage of ground-water, say fifteen feet below the surface of the ground, is healthy ; a persistently high stage of ground-water, less than five feet below the surface, is unhealthy, while a fluctuating level of the ground-water, especially if the changes are sudden and violent. is very unhealthy. This would lead us to expect that places where this fluctuation is very great would show a large mortality from such diseases as are attributed to impurities in the soil. And this we find especially true in India. In certain localities in India, cholera, for example, is endemic—that is to say, the disease is never entirely absent in such localities. Calcutta is one of these places. The rainy season begins about the first of May and continues until the end of October. During the next six months there is very little rain. It is fair to assume that the ground-water rises during the rainy season and checks decay and the multiplication of the germs of the disease in the soil, and that these processes become more active as the dry season advances and the ground-water level falls. If we note the death-rate from cholera in Calcutta it will be found that it bears a distinct relation to the movement of the ground-water. The deaths from cholera begin to increase from October and reach their height in April. Dr. Macpherson, who has written a very elaborate history of Asiatic cholera, shows this relation very clearly. For twenty-six years the average rainfall was sixty-three inches. From May to October fifty-seven inches fell, while the remaining six inches fell from November to April. The average number of deaths from cholera annually was four thousand and thirteen. Of these, one thousand two hundred and thirty-eight died in the rainy season, while two thousand seven hundred and seventy-five, nearly three-fourths, died during the period of dry weather.

In the cholera epidemics of 1866 and 1873 in Budapesth, the same relations existed between the ground-water and the cholera. As the level of the ground-water rose the cholera diminished, while the disease increased upon the sinking of the ground-water. Exactly the same behavior was exhibited by the disease in Munich in 1873.

There seems good reason to believe that typhoid fever is propagated in consequence of movements of the ground-water, in the same way as above explained for cholera. This does not exclude the infection of drinking water by the disease-germ, since much of the drinking water used, as before stated, is drawn from the ground-water. Pettenkofer, Buhl and Virchow, have shown that the death-rate from typhoid fever has a distinct and definite relation to the ground-water oscillations. This has been incontestably proven for two cities, Munich and Berlin. When the level of the ground-water is above the average, typhoid fever decreases; when it is below the average, the number of cases becomes greater. It has been mathematically demonstrated that in Munich the probability of the coincidence of a low stage of the ground-water with an increase of typhoid fever is in the ratio of 36,000 to 1. Hence, it may be regarded as an established law that the rise and fall of the ground-water bears a definite relation to the morbility rate of typhoid fever.

About twenty years ago Dr. Henry I. Bowditch, of Boston, called attention to the frequent connexion between cases of pulmonary consumption and dampness of the soil upon which the patients lived. After a very extended and laborious investigation Dr. Bowditch formulated these two propositions:

'First—A residence in or near a damp soil, whether that dampness be inherent in the soil itself or caused by percolation from adjacent ponds, rivers, meadows, or springy soils, is one of the principal causes of consumption in Massachusetts, probably in New England, and possibly other portions of the globe.

'Second—Consumption can be checked in its career, and possibly—nay, probably—prevented in some instances by attention to this law.'*

Dr. Buchanan, of England, about the same time showed that the thorough drainage of certain English cities had markedly diminished the deaths from consumption in the drained cities. So far as the writer is aware not a single fact has been established which militates against the law laid down by Dr. Bowditch and so strongly supported by the statistical researches of Dr. Buchanan, yet hardly any notice has been taken of these results by physicians. Few know anything of them and still fewer seem to have made practical use of such knowledge in advising patients. As corroborative of the views of Dr. Bowditch the rarity of consumption in high and dry mountainous districts or plateaus may be cited.

DISEASES OF ANIMALS PROBABLY DUE TO SIMILAR CONDITIONS OF THE SOIL.

The modern study of the sanitary relations of the soil is still in its infancy. Whatever definite knowl-

* Consumption in New England and elsewhere. Second Edition, Boston, 1866, p. 67.

edge has been gained relates merely to physical or chemical conditions of the soil and its atmosphere and moisture, or possibly the relations of these to the spread of certain diseases in human beings. But there is perhaps, a wider application that may be made of such knowledge than has been heretofore suggested. The domestic animals which form such a large portion of the wealth of this country—horses, cattle, sheep and hogs—are liable to infectious and contagious diseases, as well as are human beings, and many millions of dollars are lost annually by the ravages of such diseases. Now, from what is known of such diseases as *splenic fever* among cattle, and of the so-called *swine-plague*, it does not appear improbable to the writer that the source of infection is a soil polluted by the poisonous germ of these diseases, just as it seems demonstrated that cholera and typhoid fever, and possibly malarial fevers are so caused. The laborious investigations of M. Pasteur in France have shown that the cause of splenic fever when once introduced into a locality will remain active for months and even years, and it seems probable that a study of the soil in its relations to the diseases of domestic animals is a subject to which attention may profitably be given.

It is well-known that milch-cows frequently suffer from a disease identical in its nature with consumption in human beings. It is believed by many that the milk of such animals is not only unfit for food by reason of its poor quality, but that it may convey the disease to human beings when used as food. The observations of Bowditch and Buchanan, quoted above, show that consumption in man may be, and doubtless is frequently caused by soil-wetness. It seems probable that the same cause should produce similar effects in the lower animals, and it is the writer's firm conviction that an examination into the circumstances under which cows become attacked by consumption would prove this probability a fact.

DRAINAGE.

In many soils drainage is necessary in order to secure a constant level of the ground-water at a sufficient depth below the surface. Drainage and sewerage must not be confounded with each other. Drainage contemplates only the removal of the ground-water, or the reduction of its level, while sewerage aims to remove the refuse from dwellings and manufactories, including excrementitious matters, waste water and other products, and in some cases the storm water.

Sewers should never be used as drains, although for economy's sake, sewer and drainage pipes may be laid in the same trench. Sewer pipe must be perfectly air and water tight to prevent escape of its liquid or gaseous contents into the surrounding soil and rendering it impure. Drainage pipe, on the other hand, should be porous and admit water freely from without. Escape of the contents of the drain pipe into the surrounding soil will not produce any pollution of the latter.

The best material for drains is porous earthenware pipe, or the ordinary agricultural drain-tile. Coarse gravel, or broken stones may also be used, and prove efficient if the drains are properly constructed. Referring again to the aphorism of Prof. De Chaumont, that a persistently low ground-water, say fifteen feet down, or more, is healthy; that a persistently high ground-water, less than five feet from the surface is unhealthy; and that a fluctuating level, especially if the changes are sudden and violent, is very unhealthy, the necessity appears obvious, that in the construction of drainage works, the drains should be placed at a sufficient depth to secure a level of the ground-water consistent with health.   This depth should never be less than ten feet and if possible, not less than fifteen feet.   Care must be taken that the outflow of the drain is unobstructed, in order that the soil may be kept properly dry at all times.

In the absence of a proper mechanical system of drainage, the planting of certain trees may efficiently drain the soil.   It has been found that the Eucalyptus tree has produced drying of the soil when planted in sufficient numbers in marshy land.   The roots absorb a prodigious quantity of water which is then given off by evaporation from the leaves.   Sunflower plants have a similar effect upon wet soils.

CHAPTER V

In all larger communities, certain arrangements are necessary to secure a prompt and efficient removal of excreta and the refuse and used water of households and manufacturing establishments, the sweepings of streets, and rain water.

The total quantity of excrementitious products, feces and urine, for each individual, including men, women and children, has been estimated by Dr. Parkes as two and a half ounces of fecal, and forty ounces of urinary discharge daily. This would give for a population of 1,000 persons, 25 tons of feces and 91,250 gallons of urine per year. If to this is added a minimum allowance of thirty-five gallons of water per day to each individual, a complete sewerage system for a population of 1,000 persons would require provision for the discharge of 35,279½ gallons of sewage passing through the sewers every day. In this estimate, storm water, and such accessary feeders of the sewage are omitted.

The organic matters contained in sewage, even if free from the specific germs of disease, give rise to noxious emanations, which, when inhaled, probably produce a gradual depravement of nutrition, that renders the system an easier prey to disease. For this, and other reasons, it is important that such measures be adopted as will secure the removal of sewage matters from the immediate vicinage of houses as quickly as possible after they have been discharged.

The impregnation of the soil with sewage produces a contamination of ground-air and ground-water, which may become a source of grave danger to health. By polluting the ground-water, it eventually vitiates the well water which is nearly always derived from that source.

The system of removal of excrementitious matters, which any community will adopt, depends to a considerable extent upon financial considerations. Although the sanitarian must insist upon the pre-eminent importance of the cause of public health, his suggestions will receive little attention from municipal or state legislatures unless they can be carried out without involving the community too deeply in debt. For this reason it is a matter of great practical importance that the student of sanitary science should make himself familiar with the relative cost as well as with the hygienic significance of the various methods of sewage removal in use.

The different systems in use for the removal of sewage matters may be considered in detail under the following five heads :

1. The common privy, or privy vault systems.

2. The Rochdale or pail system, and its modifications.

3. The earth or ash closet system.

4. The pneumatic system of Liernur.

5. The water carriage systems.

1. *The Privy and Privy-well systems.*—While from a sanitary point of view, privies of all kinds, whether wells or cess-pits, are to be unreservedly condemned, it is not likely that they will cease to be built for many years to come. It becomes necessary therefore, to point out by what means the objections against them may be diminished, and their evil consequences, in some measure averted.

In the first place, a privy vault should be perfectly water-tight, in order to prevent pollution of the surrounding soil by transudation of the contained excremental matters. The walls should be of hard-burned brick, laid in cement. The cavity should be small in order that the contents may be frequently removed, and not allowed to remain and putrefy for months or years. A water-tight hogshead sunk in the ground makes an economical privy tank or receiver. A privy must not be dug in a cellar, or in too close proximity to the house-walls. Unless these last precautions are taken the offensive gases from the mass of decomposing fecal matter in the privy will constantly ascend into, and permeate the air of the house.

All privies should be ventilated by a pipe passing from just under the privy seat to a height of some feet above the roof of the house. A gas flame, kept burning in the upper portion of this pipe will increase its ventilating power, by creating a strong and constant upward current.

Disinfection of the contents of privies may be secured in a measure by means of sulphate of iron, carbolic acid, or dry earth. The first named is probably the most economical, most easily applied and most effective. A solution containing from one to two pounds of the salt in a gallon of water, is poured into the privy as often as necessary to prevent offensive odors. This solution may be conveniently prepared by suspending a basket or bag containing about 60 pounds of the sulphate in a barrel of water. In this way a saturated solution will be maintained until the salt has been entirely dissolved.

The most rigid disinfection by chemicals will however be less effective than thorough ventilation, for it must be remembered that the mere destruction of an offensive odor is not equivalent to removing all the deleterious properties that may be present. It is not at all certain that those elements of sewage which are the most offensive to the sense of smell are most detrimental to health.

Privies should be emptied of their contents at
stated intervals. A strict supervision should be
exercised over them by the municipal authorities in
cities and towns to prevent overflowing of their contents.

In many places the method of removing the
contents of privies is the primitive one with shovel, or
dipper and bucket. In most cities and large towns
however, the privy vaults or tanks are now emptied by
means of one of the so-called odorless excavating
machines, of which there are a number of different
patents. The process is rarely entirely odorless however, as the carelessness of the workmen frequently
permits offensive gases to escape and pollute the air for
a considerable distance. All the different forms of the
apparatus act upon the pneumatic principle. One end
of a large tube is carried into the cess-pool or vault to
be emptied, and the other attached to a pump, by means
of which the material is pumped into a strong barrel-
tank carried on wheels. At the top of the tank is a
vent, over which is placed a small charcoal furnace to
consume the foul gases escaping from the vent.

In some cities, and many of the smaller towns and
villages in this country, the primitive midden or pit
system is still in use. A shallow pit is dug in the
ground, over which is erected the privy. When the
pit is full, another is dug close by the side of it, and
the earth from the new pit thrown upon the excrement
in the old one. The privy is then moved over the new
pit and this is used until it too becomes full. The
proceeding is repeated as often as the pit becomes filled
up with the excreta, until in the course of a few years
all the available space in a yard has been honey-combed
with the pits. Then the custom adopted in overcrowded cemeteries is followed; namely, the first pit is
dug out again, and the cycle is repeated.

In other cities the privy-well system is largely in use. This is—next to the midden or shallow pit just described—the most pernicious system for the disposal of excreta that can be imagined. The wells are dug to such a depth as to reach the subterranean flow of water, in which the excremental matters are constantly carried off. Hence these receptacles never fill up, and never need cleaning. For this reason they are popular with property owners; for next to the primitive midden they are the most economical of all the various methods adopted. The utter perniciousness of the system is however, plain, because the soil for a considerable distance around each of these wells becomes a mass of putrid filth, contaminating the ground-water which feeds the drinking water supplies in the vicinity; polluting also the ground-air, which eventually reaches the surface, or the interior of houses, when the pressure of the outside atmosphere diminishes, or the ground-water level rises. It must therefore be evident that the best ventilating arrangements, or the most thorough and consistent disinfection can have very little, if any, effect in removing the very grave objections to this baneful system.

The privy-well system for the removal of excreta cannot be recommended for adoption by any sanitarian.

2. *The Rochdale, or Pail-closet System.*—The Rochdale system of removal of excreta has won the support of many distinguished sanitarians, on account of its simplicity, its economy, and its compliance with most sanitary requirements. The excreta, both solid and liquid, are received into a water-tight pail, either of wood or metal, and removed once or oftener a week; a clean and disinfected pail being substituted for the one removed. In Rochdale, Manchester and Glasgow in Great Britain, in Heidelberg in Germany, and in other cities abroad, where the system has been introduced, it

has worked satisfactorily. In this country a modification of the pail system known as the Eagle Sanitary Closet, has been introduced by a firm in Charleston, S. C. The receptacle consists of an enameled iron reservoir, with a neck just large enough to fit under the seat of the privy, and a quantity of disinfectant solution is put into the receptacle, to prevent putrefaction of the excreta. The receptacles are replaced by clean ones every week.

Mr. Jas. T. Gardner, Director of the N. Y. State Sanitary Survey, says in a special report on methods of sewerage applicable in small towns and villages, concerning the pail system :*

'Rochdale is a city of some 70,000, and Manchester of between 400,000 and 500,000 inhabitants. The higher class of houses are allowed to have water-closets, but four-fifths of the people are obliged to have "pail closets" in their yards built according to plans of the Health Department. Their essential features are: A flag-stone floor raised a few inches above the level of the yard; a hinged seat with a metal rim underneath for directing urine into the pail, which stands on the flag directly beneath the seat; a hinged front and back to the seat so that the pail or tub may be easily taken out and the place cleaned; and a six-inch ventilating pipe from under the seat to above the roof. In Rochdale they use a wooden pail or tub made of half of a disused paraffine cask holding about 100 pounds; in Manchester the "pail" is of galvanized iron and holds ten gallons. Under the direction of the authorities they are removed once a week in covered vans, which bring clean tubs to be put in the place of the full ones taken away. Each tub is covered with a close fitting double lid before removal. The tubs are taken to a depot, where their contents are deodorised and pre-

---

* Second Annual Report of New York State Board of Health, p. 322-3.

pared as manure by mixing with ashes and a small proportion of gypsum to fix the ammonia. Subsequently
street sweepings and the refuse of slaughter-houses are
added. At Manchester there is by the side of each
closet a very simple ash sifter, from which the ashes
fall into the tub and help to deodorise its contents.

'The manure at Rochdale sells for about four-fifths
of the cost of the collection and preparation.

'In 1873 the net cost to the town of removing and
disposing of the house dry refuse and excrement was
only about $95 per annum per 1,000 of population; less
than ten cents a person per annum.

'The system has been in operation more than
twelve years.

'The tubs are removed in the day time without
offensive odor.

'Where ashes are frequently thrown into the tubs
at Manchester, very little odor is to be perceived in the
closets.

'For the villages of the State which can have no
general water supply, I would unhesitatingly advise
the use of the "pail" or tub system as practised in
Manchester, England, as being, from a sanitary point
of view, an immense improvement over the death-
breeding *privy-vaults* in common use. The cheapness
of the plan and the smallness of the original outlay of
brains and money, in comparison with that needed to
build a good sewer system, will make it possible to in-
troduce a tub-privy system into most villages half a
century before sewers would meet with any consideration.

'At a small cost the existing privy-vaults can be
cleaned and filled, and the privies altered into healthful
tub-closets. The town authorities must then arrange
for the removal of the tubs once a week, and for their
thorough cleansing and disinfecting. Any isolated

house or group of houses can use the tub system, taking
care of it themselves. If the plan is adopted in villages, it will doubtless spread into the country, and
become the most powerful means of abolishing the fatal
privy-vaults which are poisoning the farm wells.'

3. *Earth and Ash Closets.*—The earth and ash
closets are devices in use to a large extent in England,
and to a less degree in this country, for the purpose of
rendering human excreta inodorous by covering them
immediately after they are voided with dry earth or
ashes. The earth-closet is the invention of the Rev.
Henry Moule, of England, and consists of an ordinary
commode or closet, the essential feature of which is a
reservoir containing dried earth or ashes, a quantity
of which, amounting to about twice the quantity of
feces voided, is thrown upon the evacuation either by
hand, or by means of an automatic apparatus called a
'chucker.' Just as in the ordinary water-closet, by
raising a handle a supply of water is thrown into the
hopper to wash down the feces into the soil-pipe, so in
the usual form of the earth-closet, raising the handle
projects a quantity of earth upon the evacuated feces
and urine. By this means the excreta are rendered
entirely inodorous and dry. The contents of the closets
may be collected into a heap in a dry place. In the
course of a few months the organic constituents have
become oxidised and the earth may be used over again,
for a number of times. A well-known sanitarian states
that he has used sifted anthracite coal ashes ten or
twelve times over, in the course of three years. During
this time the material under no circumstances gave any
indication that it was 'anything but ashes, with
a slight admixture of garden soil.'*

Dr. Buchanan, of England, comparing the advantages of the earth-closet with those of the water-closet,

---

* WARING; The Sanitary Drainage of Houses and Towns. 2nd Ed., 1881, p. 250.

says: 'It is cheaper in original cost; it requires less repairs; it is not injured by frost; it is not damaged by improper substances being thrown down it; and it very greatly reduces the quantity of water required by each household.'*

In cities and towns, the removal of the excreta should be carried out by, or under the immediate direction of the municipal sanitary authorities. If this is neglected, abuses are liable to creep in which will vitiate the performance of any system however faultless if properly managed.

Many advocates of the pail, dry earth or privy-systems, urge the advantage of the large quantity of valuable manure which can be realised by converting the excremental matters into poudrette and other fertilising compounds. Experience has shown, however, that the cost of preparing a satisfactory fertiliser from human excrement is much greater than can be realised from its sale. In all places in Great Britain and the continent of Europe where it has been tried, the decision is against its practicability. The agricultural consideration should, however, be a secondary one, if the systems mentioned are economical and meet the sanitary requirements (which the privy-system certainly does not). The adoption of one, or other of them may be secured, where more perfect, but more complicated and expensive systems may be out of the question.

4. *The Pneumatic System of Liernur.*—A system which seems to be useful in larger cities, especially where the topographical conditions are such as to render necessary mechanical aid in overcoming obstacles to natural drainage, is the pneumatic system devised by Capt. Liernur of Holland, and generally known as the Liernur system. It consists of a set of soil pipes running from the water-closets to central

---

* Quoted in WARING, above cited, p. 264.

district reservoirs, from which the air is exhausted at
stated intervals. When a vacuum is created in the
reservoir the contents of the water-closets and soil
pipes are driven forcibly into the reservoir by the
pressure of air. The district reservoirs are connected
by a separate system of pipes with a main depot and,
the transfer of the fecal matter from the former to
the latter is also accomplished with the aid of pneumatic pressure. The complete system of Liernur pro-
vides that at the main depot, the fecal matter shall be
treated with chemicals, evaporated, and converted into a
dry fertiliser—poudrette. It appears from the published reports that while the system has been partially
adopted in three Dutch cities, in only one of them,
Dortrecht, has the machinery for manufacturing poudrette been established. With reference to this, Erismann* says: 'It seems never to have been in regular
working order, for the fecal masses are mixed with
street-sweepings and ashes into a compost-mass, which
causes no little discomfort in the neighborhood, by the
offensive odors. In Amsterdam, the fecal matters,
which frequently do not find a ready sale, are partly
made into a compost with sweepings, partly used to
fertilize meadows, or simply discharged into the water.

As to the practical working of the system, the
opinions differ widely. While the majority of sanitarians including Virchow, von Pettenkofer, and Mr.
Rawlinson, object to it as not fulfilling the demands of
hygiene, the system has also been criticised by engineers
as not being in accordance with the well-known principles of their science.†

---

* Von Pettenkofer und Ziemssen. Handbuch der Hygiene. II Th. II Abth, 1
Hlfte. p. 140.

† Papers by Maj. C. H. Latrobe and Col. Geo. E. Waring, Jr., in Fifth Bien-
nial Report, Md. State Board of Health. See also, in favor of system, a paper by
Dr. C. W. Chancellor in same publication and an elaborate description by the same
author in Trans. Med. and Chir. Faculty of Md., 1883.

Two other plans for the removal of fecal matter by pneumatic pressure have been invented, namely the Shone and the Berlier systems. Neither of these has been adopted to any extent. Both seem to the author to fall far short even of the merits of the Liernur system.

5. *The Water-Carriage System of Sewerage.*—Two systems of removal of sewage by water-carriage are in use at the present time. They are technically known as the 'combined' and the 'separate' systems. In the former, which is the system upon which most of the sewers in this country are constructed, all excreta, kitchen slops, waste water from baths and manufacturing establishments, as well as storm water are carried off in the same conduits. In the separate system, on the other hand, the removal of the storm water is provided for, either by surface or underground drains, not connected with the sewers proper, in which only the discharge from water-closets and the refuse water from houses and factories are conveyed. In the separate system, the pipes are of such small calibre that a constant flow of their contents is maintained, preventing deposition of suspended matters, and diminishing decomposition and the formation of sewer gas.

In the combined system, on the other hand, the sewers must be made large enough to receive the maximum rainfall of the district. This requires a calibre greatly in excess of the ordinary needs of the sewer, and furnishes favorable conditions for the formation of sewer gas, and the development of minute vegetable organisms. The ordinary flow in a sewer of large calibre is usually so sluggish as to promote the deposition of solid matters and gradual obstruction of the sewer.

It is the opinion of the most advanced sanitarians, that the separate system fulfils the demands of a rational system of sewerage better than any other at present in use. The objections to the combined system are so many and so great, that it does not seem advisable for sanitary authorities to recommend the construction of sewers on this principle in the future.

The separate system of sewerage endorsed as it is by high engineering and sanitary authorities, and by a satisfactory, practical test of four years in the city of Memphis, seems to the author to possess merits above any other plan for the removal of excreta and house wastes. The following description is from a paper by Col. George E. Waring, Jr.: 'A perfect system of sanitary sewerage would be something like the following: No sewer should be used of a smaller diameter than six inches: *a*, because it will not be safe to adopt a smaller size than four-inch for house drains, and the sewer must be large enough to surely remove whatever may be delivered by these: *b*, because a smaller pipe than six-inch would be less readily ventilated than is desirable: *c*, and because it is not necessary to adopt a smaller radius than three inches to secure a cleansing of the channel by reasonably copious flushing.

'No sewer should be more than six inches in diameter until it and its branches have accumulated a sufficient flow at the hour of greatest use to fill this size half full, because the use of a larger size would be wasteful, and because when a sufficient ventilating capacity is secured, as it is in the use of a six-inch pipe, the ventilation becomes less complete as the size increases—leaving a larger volume of contained air to be moved by the friction of the current, or by extraneous influences, or to be acted upon by changes of temperature, and of volume of flow within the sewer.

'The size should be increased gradually and only so rapidly as is made necessary by the filling of the sewer half full at the hour of greatest flow.

'Every point of the sewer should, by the use of gaskets or otherwise, be protected against the least intrusion of cement, which, in spite of the greatest care, creates a roughness that is liable to accumulate obstructions.

'The upper end of each branch sewer should be provided with a Field's flush tank of sufficient capacity to secure the thorough daily cleansing of so much of the conduit as from its limited flow is liable to deposit solid matters by the way.

'There should be sufficient man-holes, covered by open gratings, to admit air for ventilation. If the directions already given are adhered to, man-holes will not be necessary for cleansing. The use of the flush-tank will be a safeguard against deposit. With the system of ventilation about to be described, it will suffice to place the man-holes at intervals of not less than 1,000 feet.

'For the complete ventilation of the sewers it should be made compulsory for every householder to make his connexion without a trap, and to continue his soil-pipe above the roof of his house. That is, every house connexion should furnish an uninterrupted ventilation channel four inches in diameter throughout its entire length. This is directly the reverse of the system of connexion that should be adopted in the case of storm-water and street-wash sewers. These are foul, and the volume of their contained air is too great to be thoroughly ventilated by such appliances. Their atmosphere contains too much of the impure gases to make it prudent to discharge it through house-drains and soil-pipes. With the system of small pipes now described, the flushing would be so constant and complete, and the amount of ventilation furnished, as compared to volume of air to be changed, would be so great, that what is popularly known as "sewer-gas" would never exist in any part

of the public drains. Even the gases produced in the
traps and pipes of the house itself would be amply
rectified, diluted, and removed by the constant movement of air through the latter.

'All house connexions with the sewers should be
through inlets entering in the direction of the flow,
and these inlets should be funnel-shaped so that their
flow may be delivered at the bottom of the sewer, and
so that they may withdraw the air from its crown; that
is, the vertical diameter of the inlet at its point of
junction should be the same as the diameter of the
sewer.

'All changes of direction should be on gradual
curves, and, as a matter of course, the fall from the
head of each branch to the outlet should be continuous.
Reduction of grade within this limit, if considerable,
should always be gradual.

'So far as circumstances will allow, the drains
should be brought together, and they should finally discharge through one or a few main outlets.

'The outlet, if water-locked, should have ample
means for the admission of fresh air. If open, the
mouth should be protected against the direct action of
the wind.

'It will be seen that the system of sewerage here
described is radically different from the usual practice.
It is cleaner, is much more completely ventilated, and
is more exactly suited to the work to be performed.
It obviates the filthy accumulation of street manure in
catch-basins and sewers, and it discharges all that is
delivered to it at the point of ultimate outlet outside
the town before decomposition can even begin. If
the discharge is of domestic sewage only, its solid matter will be consumed by fishes if it is delivered into a
water-course, and its dissolved material will be taken
up by aquatic vegetation.

'The limited quantity and the uniform volume of the sewage, together with the absence of dilution by rain-fall, will make its disposal by agricultural or chemical processes easy and reliable.

'The cost of construction, as compared with that of the most restricted storm-water sewers, will be so small as to bring the improvement within the reach of the smaller communities.

'In other words, while the system is the best for large cities, it is the only one that can be afforded in the case of small towns.

'Circumstances are occasionally such as to require extensive engineering works for the removal of storm-water through very deep channels. Ordinarily, the removal of storm-water is a very simple matter, if we will accept the fact that it is best carried, so far as possible, by surface gutters, or, in certain cases by special conduits, placed near the surface.

'It is often necessary, in addition to the removal of house waste, to provide for the drainage of the sub-soil. This should not be effected by open joints in the sewers: because the same opening that admits soil-water may, in dry seasons, and porous soils, permit the escape of sewage matters into the ground, which is always objectionable.

'Soil-water drains may be laid in the same trench with the sewers, but preferably, unless they have an independent outlet, on a shelf at a higher level. When they discharge into the sewer, they should always deliver into its uppper part, or into a man-hole at a point above the flow-line of the sewage.'*

The establishment of a system of sewerage presupposes a constant and abundant supply of water to keep all closets clean, and all house-drains and street sewers well flushed. Where this cannot be obtained,

---

* WARING: The Sewering and Drainage of Cities. Public Health. Vol. V., p. 35.

sewers would be likely to prove greater evils than ben-
efits. In such cases one of the methods of removal of
excreta above mentioned, either the pail or earth-closet
system should be adopted.

The final disposal of sewage is a problem that depends for its solution partly upon the agricultural
needs of the country around the city to be sewered,
partly upon the proximity of large bodies of water or
running streams. When the city is situated upon, or
near large and swiftly flowing streams, the sewage
may be emptied directly into the stream without seriously impairing the purity of the latter. Dilution,
deposition and oxidation will soon remove all appreciable traces of the sewage of even the largest cities.
Where, on the other hand, the stream is inadequate in
size to carry off the sewage, or where, as in the Seine
and Thames, the current is sluggish, some other method
of final disposal must be adopted.

In many cities of Great Britain and the continent
of Europe the disposal of the sewage by irrigation of
cultivated land has been practised for a number of
years. The reports upon the working of the system are
generally favorable, although some sanitarians express
doubts of the efficiency of the system. In using sewage
for the irrigation of land, two objects are secured, first,
the fertilisation of the land by the manurial constituents of the sewage, and second, the purification of the
liquid portion by filtration through the soil. The organic matters which have been held back by the soil
undergo rapid oxidation in the presence of air and the
bacteria of decay, and are converted into plant-food,
or into harmless compounds.

Sewage irrigation as practised in Europe, must
make provision for the disposal of a very large proportion
of water in the sewage (street-wash, storm-water) which
requires much larger areas of land than would be needed if

only sewage material proper (water-closet and kitchen waste) was to be thus disposed of. In this country a practical experiment has recently been made with a well constructed separate system of sewerage, delivering only the sewage materials above mentioned upon the irrigation area.* The success of the experiment is said to be exceedingly satisfactory.

All land used for sewage irrigation should be drained with drain tile at a depth of three to six feet below the surface, in order to promote a rapid carrying off of the watery portion of the sewage, purified by filtration through the soil. A sandy loam is the best soil for irrigation. Clay is not sufficiently permeable to air and water, while pure sand allows the sewage to pass through too readily, before the organic matters in it have been sufficiently oxidised.

It has been shown that the roots of plants assist largely in the oxidation of organic matter.

The entire process of collecting and finally disposing of sewage matters from the moment they are received in the house receptacles until discharged into the swiftly flowing stream or on the sewage farm, should be void of offense to the senses of sight or smell. With a proper construction and management of sewerage works, on the lines indicated in this chapter, it is believed these results can be attained.

[The following works give fuller details upon the matters treated in the two foregoing chapters:

ERISMANN: Entfernung der Abfallstoffe. Hdbch d. Hygiene, etc. II Th. I. Abth. I. Hlfte. C. F. FOLSOM. The Disposal of Sewage, Seventh Rep't Mass. State Board of Health , 1876, p. 276. SOYKA: Städte-reinigung, in Realencyclopædie d. ges. Heilk. Bd. XIII., p. 14, et. seq. W. H. FORD. Soil and Water, in BUCK's Hygiene and Public Health, Vol. I. PETTENKOFER: The Sanitary Relations of 'the Soil. Pop. Sci. Monthly, Vol, XX, p· 332, 468.]

---

* Pullman from a State Medicine Point of View. By O. C. DE WOLF, M. D. Public Health, Vol. IX., p. 290.

CHAPTER VI.

CONSTRUCTION OF HABITATIONS.

THE importance of observing the principles of hygiene in the construction of habitations for human beings is not sufficiently appreciated by the public. Architects and builders themselves have not kept pace with the sanitarian, in the study of the conditions necessary to be observed in building a dwelling house which shall answer the requirements of sanitary science.

In an investigation conducted by Dr. Villermé* it was found that in France, from 1821–1827, of the inhabitants of arrondissements containing 7 per cent. of badly constructed dwellings, one person out of every seventy-two died. Of inhabitants of arrondissements containing 22 per cent. of badly constructed dwellings, one out of sixty-five died, while of the inhabitants of arrondissements containing 38 per cent. of badly constructed dwellings, one out of every forty-five died.

Inseparable from the question of the defective construction of dwellings, is that of overcrowding in cities, because the most crowded portions of a city are at the same time those in which the construction of dwellings is most defective from a hygienic standpoint. The following tables show the relations of the death-rate to density of population in various large cities of Europe; and also the relations between overcrowding in dwellings and the mortality from contagious diseases:

Relation of Death-rate to Density of Population.

CITY.	MEAN NUMBER OF INHABITANTS TO EACH HOUSE.	AVERAGE DEATH-RATE PER 1,000 INHABITANTS.
London,	8	24
Berlin,	32	25
Paris,	35	28
St. Petersburg,	52	41
Vienna,	55	47

* Quoted in Realencyclopædia d. ges. Heilk, Bd. II., 71.

In Berlin in 1872-3 it was found that out of every one hundred deaths from all causes, there were from contagious diseases:

20 Deaths in Dwellings with 1-2 Persons in each Room.
20   "   "   "   " 3-5   "   "   "   "
32   "   "   "   " (5-10)   "   "   "   "
79   "   "   " with over 10   "   "   "   "

These figures show very clearly the vital importance of the application of sanitary laws in the construction of dwellings.

Another curious and suggestive point is presented by some statistical researches on the mortality of Berlin, in regard to the death-rate among persons living in different stories of houses. It was found, for example, that the mortality in fourth-story dwellings is higher than in the lower stories. Even basement-dwellings furnish a lower death-rate. Still-births especially, occur in a larger proportion among the occupants of the upper stories of houses.

It is in the death-rate among young children, that the effects of overcrowding and unsanitary construction of dwellings, are especially manifest. The mortality returns from all the large cities of the world give mournful evidences of this every summer.

The researches of Dr. H. I. Bowditch upon soil-wetness, to which reference has already been made in a previous chapter, show conclusively that persons living in houses situated upon or near land habitually or excessively wet, are especially prone to be attacked by pulmonary consumption. Dr. Buchanan* has corroborated the truth of Dr. Bowditch's observations by showing from the records of a number of cities and towns of Great Britain that with the introduction of a good drainage system, bringing about a depression and uniformity of level of the ground-water, the mortality from consumption and other diseases very markedly

* Ninth and Tenth Reports of the Medical Officer to the Privy Council.

diminished.   The following table showing the propor-
tionate amount of this diminution, is abridged from
the official reports :*

Results of Sanitary Work.

NAME OF PLACE.	POPULATION IN 1861.	AVERAGE MORTALITY PER 1,000 BEFORE CONSTRUCTION OF WORKS.	AVERAGE MORTALITY PER 1,000 SINCE COMPLETION OF WORKS.	SAVING OF LIFE. PER CT.	REDUCTION OF TYPHOID FEVER RATE. PER CT.	REDUCTION IN RATE OF PHTHISIS. PER CT.
Banbury, -	10,238	23.4	20.5	12½	48	41
Cardiff, -	32,954	33.2	22.6	32	40	17
Croydon, -	30,229	23.7	18.6	22	63	17
Dover, -	23,108	22.6	20.9	7	36	20
Ely, - -	7,847	23.9	20.5	14	56	47
Leicester, -	68,056	26.4	25.2	4½	48	32
Macclesfield,	27,475	29.8	23.7	20	48	31
Merthyr, -	52,778	33.2	26.2	18	60	11
Newport, -	24,756	31.8	21.6	32	36	32
Rugby, -	7,818	19.1	18.6	2½	10	43
Salisbury, -	9,030	27.5	21.9	20	75	49
Warwick, -	10,570	22.7	21.0	7½	52	19

The following points must be taken into account in
building a house in accordance with sanitary principles :

I.—SITE.

The building site should be protected against violent winds, although a free circulation of air all around
the house must be secured.   Close proximity to cemeteries, marshes, and injurious manufacturing establishments, or industries, must be avoided, if possible.   A
requisite of the highest importance is the ability to
command an abundant supply of pure water for drinking, and other purposes.   A neglect of this precaution
will be sure to result to the serious inconvenience, if
not detriment of the occupants of the house.

II.—CHARACTER OF THE SOIL.

The soil should be porous and free from decomposing animal or vegetable remains, or excreta of man
or animals.   It should be freely permeable to air and
water, and the highest level of the ground-water should
never approach nearer than ten feet to the surface.

---

* BALDWIN LATHAM: Sanitary Engineering. Chicago, 1877, p 2.

The fluctuations of the ground-water level should be limited. In this connexion, attention is again called to the aphorism of Dr. De Chaumont.*

It is impossible to say positively that any kind of soil is either healthy or unhealthy, merely from a knowledge of its geological characters. The accidental modifying conditions above referred to, viz: organic impurities, moisture, the level and fluctuations of the ground-water are of much greater importance than mere geological formation. The late Dr. Parkes, however, regarded the geological structure and conformation as of no little importance, and summarized the sanitary relations of soils variously constituted, as follows:†

'1. *The Granitic, Metamorphic, and Trap Rocks.*— Sites on these formations are usually healthy; the slope is great, water runs off readily; the air is comparatively dry; vegetation is not excessive; marshes and malaria are comparatively infrequent, and few impurities pass into the drinking water.

'When these rocks have been weathered and disintegrated, they are supposed to be unhealthy. Such soil is absorbent of water; and the disintegrated granite of Hong Kong is said to be rapidly permeated by a fungus; but evidence as to the effect of disintegrated granite or trap is really wanting.

'In Brazil the syenite becomes coated with a dark substance, and looks like plumbago, and the Indians believe this gives rise to "calentura," or fevers. The dark granitoid, or metamorphic trap, or hornblendic rocks in Mysore, are also said to cause periodic fevers; and iron hornblende especially was affirmed by Dr. Heyne, of Madras, to be dangerous in this respect. But the observations of Richter on similar rocks in

* Chapter IV., p. 97.

† Practical Hygiene, 6th Ed., Vol. I., p. 359.

Saxony, and the fact that stations on the lower spurs of the Himalayas on such rocks are quite healthy, negative Heyne's opinion.

'2. *The Clay Slate.*—These rocks precisely resemble the granite and granitoid formations in their effect on health. They have usually much slope; are very impermeable; vegetation is scanty, and nothing is added to air or drinking-water.

'They are consequently healthy. Water, however, is often scarce, and as to the granite districts, there are swollen brooks during rain, and dry water courses at other times swelling rapidly after rains.

'3. *The Limestone, and Magnesian Limestone Rocks.*—These so far resemble the former, that there is a good deal of slope, and rapid passing off of water. Marshes, however, are more common, and may exist at great heights. In that case, the marsh is probably fed with water from some of the large cavities, which, in the course of ages become hollowed out in the limestone rocks by the carbonic acid in the rain, and form reservoirs of water.

'The drinking water is hard, sparkling and clear. Of the various kinds of limestone, the hard oolite is best, and magnesian is worst; and it is desirable not to put stations on magnesian limestone if it can be avoided.

'4. *The Chalk.*—The chalk, when mixed with clay, and permeable, forms a very healthy soil. The air is pure, and the water, though charged with calcium carbonate, is clear, sparkling, and pleasant. Goitre is not nearly so common, nor apparently calculus, as in the limestone districts.

'If the chalk be marly, it becomes impermeable, and is then often damp and cold. The lower parts of the chalk which are underlaid by gault clay, and which also receive the drainage of the parts above, are often very malarious; and in America, some of the most marshy districts are in the chalk.

'5. *The Sandstones.*—The permeable sandstones are very healthy; both soil and air are dry; the drinking water is, however, sometimes impure. If the sand be mixed with much clay, or if clay underlies a shallow sand-rock, the site is sometimes damp.

'The hard millstone grit formations are very healthy, and their conditions resemble those of granite.

'6. *Gravels* of any depth are always healthy, except when they are much below the general surface, and water rises through them. Gravel hillocks are the healthiest of all sites, and the water, which often flows out in springs near the base, being held up by the underlying clay, is very pure.

'7. *Sands.*—There are both healthy and unhealthy sands. The healthy are the pure sands, which contain no organic matter, and are of considerable depth. The air is pure, and so is often the drinking-water. Sometimes the drinking-water contains enough iron to become hard, and even chalybeate. The unhealthy sands are those which, like the subsoil of the Landes, in southwest France, are composed of silicious particles (and some iron), held together by a vegetable sediment.

'In other cases sand is unhealthy, from underlying clay or laterite near the surface, or from being so placed that water rises through its permeable soil from higher levels. Water may then be found within three or four feet of the surface; and in this case the sand is unhealthy, and often malarious. Impurities are retained in it, and effluvia traverse it.

'In a third class of cases, the sands are unhealthy because they contain soluble mineral matter. Many sands (as, for example, in the Punjab) contain magnesium carbonate and lime salts, as well as salts of the alkalies. The drinking-water may thus contain large quantities of sodium chloride, sodium carbonate, and even lime and magnesian salts and iron. Without examination of the water, it is impossible to detect these points.

·8. *Clay, Dense Marls, and Alluvial Soils generally.*—These are always regarded with suspicion. Water neither runs off nor runs through; the air is moist; marshes are common; the composition of the water varies, but it is often impure with lime and soda salts. In alluvial soils there are often alterations of thin strata of sand, and sandy, impermeable clay. Much vegetable matter is often mixed with this, and air and water are both impure.

The deltas of great rivers present these alluvial characters in the highest degree, and should not be chosen for sites. If they must be taken, only the most thorough drainage can make them healthy. It is astonishing, however, what good can be effected by the drainage of even a small area, quite insufficient to affect the general atmosphere of the place ; this shows that it is the local dampness and the effluvia which are the most hurtful.

'9. *Cultivated Soils.*—Well-cultivated soils are often healthy, nor at present has it been proved that the use of manure is hurtful. Irrigated lands, and especially rice fields, which not only give a great surface for evaporation, but also send up organic matter into the air, are hurtful. In Northern Italy, where there is a very perfect system of irrigation, the rice grounds are ordered to be kept 14 kilometers (8.7 miles) from the chief cities. 9 kilometers (5.6 miles) from the lesser cities and the forts, and 1 kilometer (1.094 yards) from the smaller towns. In the rice countries of India (and America), this point should not be overlooked.'

Where a wet, impermeable, or impure soil must, of necessity, be chosen as a building site, it should be thoroughly drained. The minimum depth at which drains are laid should be not less than five feet below the floor of the cellar or basement. Such a soil should be covered with a thick impervious layer of asphaltum or similar cement under the house, in order to prevent the aspiration of the polluted ground-air into the building.

It is a frequent custom in cities to fill in irregularities of the building-site with street-sweepings, and garbage, which always contain large quantities of decomposing organic matters. This is a gross violation of the plainest principles of hygiene. It is almost equally reprehensible to use such decaying or putrefying organic material for the purpose of grading streets or sidewalks in cities and towns.* It should be the constant endeavor of all sanitary authorities to prevent pollution of the soil as much as possible in villages, towns and cities.

Where houses are built on the declivity of a hill, the upper wall should not be built directly against the ground, as it would tend to keep the wall damp. A vacant space should be left between the wall and the ground to permit free access of air and light.

In addition to, or in default of drainage, the drying of soil can be promoted by rapidly growing plants, which absorb water from the soil and give it out to the air. The sunflower and the eucalyptus tree are the most available for this purpose.

III.—THE MATERIAL OF WHICH THE HOUSE IS BUILT.

The nature of the most appropriate building material depends upon so many collateral circumstances, that

---

* During the very fatal epidemic of yellow fever in New Orleans in 1878, it was ascertained that a contractor for street-work used the garbage and street-scrapings to grade the bed of the streets. Even though in this case it may not have intensified the epidemic in these localities, the practice is so contrary to the simplest sanitary laws, that it should nowhere be tolerated. The author is aware, however, that the 'made ground' of nearly every city in this country is composed largely of just such material. All sanitarians should protest against a continuance of this pernicious practice.

definite rules cannot be laid down. As a general rule, moderately hard-burned brick is the most serviceable and available material. It is easily permeable by the air, and so permits natural ventilation through the walls, unless this is prevented by other means. It does not absorb and hold water readily, hence, damp walls are infrequent, if brick is used. It is probably of all building material the most durable. On account of its porosity, a brick wall is a poor conductor of heat. It therefore prevents the rapid cooling of a room in cold weather, and likewise retards the heating of the inside air from without, in summer. Another very great advantage, is its resistance to a very high degree of heat, brick being probably more nearly fire-proof, than any other building material.

In hot climates, light wooden buildings are advantageous, because they cool off very rapidly after the sun has disappeared. On account of the numerous joints and fissures in a frame building, natural ventilation goes on very readily and to a considerable extent.

Next to brick, granite, marble and sandstone are the most serviceable building materials. Very porous sandstone is, however, not very durable in cold climates, as the stone absorbs large quantities of water, which in consequence of the expansion accompanying the act of freezing, produces a gradual but progressive disintegration.

The application of paint to the walls, either within or without, almost completely checks the transpiration of air through the walls, thus limiting natural ventilation. Calcimining, on the other hand, offers very little obstruction to the passage of air. Wall-paper is about mid-way between paint and lime coating in its obstructive effect on atmospheric transpiration.

Newly-built houses should not be occupied until the walls have become dry. Moisture in the walls is probably a not infrequent source of ill-health ; it offers favorable conditions for the development of fungi (possibly of disease-germs), and, by filling up the pores of the material of which the walls are composed, prevents the free transpiration of air through them.

Moisture of the walls is sometimes due to ascent of the water from the soil by capillary attraction. This can be prevented by interposing an impervious layer of slate in the foundation wall.

Where the moisture is due to the rain beating against the outside walls and thus saturating them if composed of porous materials, a thorough external coating of impervious paint will prove a good remedy.

IV.—INTERIOR ARRANGEMENTS.

A. Size of Rooms, and Ventilating and Heating Arrangements.—The rooms in dwelling-houses should never be under eight feet in height from floor to ceiling. In sleeping rooms, the initial air-space should never be less than 1,200 cubic feet for adults, and 800-900 cubic feet for children under ten years of age. Provision must be made for changing this air sufficiently often to maintain it at its standard of purity; *i. e.*, less than seven parts of carbonic acid per 10,000. The details for accomplishing this will vary with the architects' designs, the material of which the house is constructed, the climate, and the season. The principles laid down in the section on ventilation, (Chap. I) should be adhered to. In cold weather, the air should be warmed either before its entrance into the room, or afterward by stove or fireplace. The details of the heating apparatus must be left to individual taste, or other circumstances. It may be noted, however, in passing, that the prevailing method of heating houses by means of hot air is objectionable for various reasons; partly, because the air is usually too dry to be comfortable to

the respiratory organs; partly, because organic matter
is frequently present in large proportions, and gives
the air an offensive odor when the degree of heat is
high enough to scorch the organic matter.   Both these
objections are however, removable; the first by keeping
a vessel of water constantly in the furnace, so that the
hot air can take up a sufficient proportion of vapor in
passing through, and the second by having the furnace
made large enough so that the temperature need never
be raised to a very high degree.

   B. Internal Wall Coating.—A point of considerable importance in the out-fitting of dwelling-houses is
the material used for coating or decorating the inside
of the walls.   Green paint, or green-colored wall papers
should be rejected.   The reason for avoiding this color
is the following.   Bright-green pigments and dyes are
largely composed of some compound of arsenic, which
becomes detached from the wall or paper when dry,
and being inhaled produces a train of symptoms, which
have been recognized as chronic arsenical poisoning.
Many cases have been reported in which serious and
even fatal poisoning has been produced in this way.*
It would be advisable, therefore, to discard all bright-
green tints in paints and ornamental paper hangings.

   C. Lighting.—Provision should be made, in all
dwelling-houses for an abundant supply of sun-light.
Every room should have at least one window opening
directly to the sun.   It is not sufficient to give an
ample window-space, which should be in the proportion
of one to five or six of floor space, but the immediate
surroundings of the house must be taken into account.
Thus, close proximity of other buildings, or of trees
may prevent sufficient light entering a room, although
the window-space may be in excess of that required
under ordinary circumstances.

---

* F. W. DRAPER: Arsenic in Certain Green Colors. Third Annual Rep't Mass.
State Board of Health, 1872. p. 18-57.

Some form of artificial light will also be needed in all dwellings. Certain dangers are necessary accompaniments of all available methods of artificial illumination. The danger from fire, is, of course, the most serious. This danger is probably least where candles are used, and greatest where the more volatile oils (kerosene, gasoline,) are employed. The use of candles results in pollution of the air by carbonic acid and other products of combustion to a greater degree than when other illuminating agents are used; they also give out a larger amount of heat in proportion to their power of illumination. Kerosene gives a good light, when burned in a proper lamp, and is cheap, but the dangers from explosion and fire are considerable. The danger from explosion can be greatly reduced by always keeping the lamp filled nearly to the top, and never filling it near a light or fire. The danger of explosion is increased when the chimney of the lamp is broken, as then the temperature of the metal collar by which the burner is fastened to the lamp is rapidly raised,* and the oil vaporized. If at the same time the lamp is only partially filled with oil, the space above it is occupied by an explosive mixture of air and the vapor of the oil. If this is heated to a sufficient degree, an explosion will take place.†

The use of coal-gas is probably attended by less danger than the lighter oils, but by more than other means of illumination. In addition to the dangers from fire and explosions, which are inevitable accompaniments of defects in the fixtures, the escaping gas is itself exceedingly poisonous from the large amount of carbonic oxide it contains. It is, in fact, a very frequent occurrence, in large cities, that persons are killed by

---

* H. B. BAKER, in Rep't Mich. State Board of Health. 1876.   p. XLVIII.

† See an instructive paper by Prof. R. C. KEDZIE in Rep't Mich. State Board of Health for 1877.   p. 71, et. seq.

the inhalation of gas which has escaped from the fixtures, or was allowed to escape from the burner, through
ignorance.

The electric light (Edison's incandescent system) is
probably open to less objection on the score of danger,
than any other of the illuminating systems mentioned.
There is no reliable evidence that the electric light has
any unfavorable influence on the vision, although Reg-
nault supposed it would have a bad effect upon the
ocular humors. on account of the large proportion of
the violet and ultra-violet rays it contained.   In order
to remove this objection, Bouchardat advised the wearing of yellow glasses by those compelled to use this
light for close work.   The advantages of the incandescent light, beside the brilliant white light it gives, are
that it is steady. and does not produce any heat, nor
does it pollute the air with carbonic acid and other
products of combustion.   Prof. von Pettenkofer has
recently shown experimentally that the pollution of
the air by the products of combustion is very much
greater when gas is used than where the electric light
is employed.

In writing, sewing, reading or other work requiring a constant use of accurate vision, the light, whether
natural or artificial, should fall upon the object from
above and on the left side.   Hence, windows and burners should be, at least. at the height of the shoulder
and to the left of the person using the light.

Increased ventilation facilities must be provided
where artificial light (except the electric light) is used
to any extent.   It has been calculated that for every
lighted gas-burner, 400–500 cubic feet of fresh air per
hour must be furnished in addition to the amount ordinarily required, in order to maintain the air of the
room at the standard of purity.

V.—WATER SUPPLY.

The water supply of a dwelling house should be plentiful for all requirements, and its distribution should be so arranged that the supply for every room is easily accessible. Where practicable, water taps should be placed on every floor, both for convenience, and for greater safety in case of fire. It is also a result of observation that personal habits of cleanliness increase in a direct ratio with the ease of obtaining the cleansing agent. The inmates of a house where water is obtainable with little exertion, are much more likely to be cleanly in habits, than where the water supply is deficient, or not readily procured.

VI.—HOUSE DRAINAGE.

Provision must be made for the rapid and thorough removal of waste water and excrementitious substances from the house. This is most easily and completely accomplished by well-constructed water-closets and sinks. Water-closets should, however, not be tolerated in any room occupied as a living or bed-room. It would doubtless be very much more in accordance with sanitary requirements to have all permanent water-fixtures, water-closets and bathing arrangements placed in an annex to the dwelling proper. In this way, the most serious danger from water-closets and all arrangements having a connexion with a cess-pool, or common sewer —permeation of the house by sewer-air—could be avoided.

Water-closets, however, presuppose an abundant supply of water. Unless this can be obtained and rendered available for flushing the closets, soil-pipe and house-drain, the dry-earth or pail system should be adopted. Privies should not be countenanced. Experience in several large cities of Europe has demonstrated* that the pail system can be adopted with advantage, and satisfactorily managed even in large communities.

As house drainage may be considered the first and most important link in a good sewerage system, a brief description will be here given of the details of the drainage arrangements of a dwelling house. The rapid and complete removal of all fecal and urinary discharges, lavatory and bath-wastes and kitchen slops must be provided for. For these purposes are needed, *first*, water-closets and urinals, wash-basins and bath-tubs, and kitchen or slop-sinks; *second*, a perpendicular pipe, with which the foregoing are connected, termed the soil-pipe, and *third*, a horizontal pipe or house-drain, connecting with the common cess-pool or sewer.

A. Water-Closets.—There are five classes of water-closets in general use. They are the pan, valve, plunger, hopper and washout closets.

Pan closets are those found in most old houses containing water-closet fixtures. Just under the bowl of the closet is a shallow pan containing a little water, in which the dejections are received. On raising the handle of the closet, the pan is tilted, and the water at the same time is turned on, which washes out the excrement and sends it into or through the trap between the closet and the soil-pipe. It will be readily understood that the space required for the movement of the pan, the 'container' as it is termed, is rarely thoroughly cleansed by the passage of water through it. Fecal matter, paper, etc., gradually accumulates in the corners of the container, and as a consequence, pan closets are always, after a brief period of use, foul. There are other defects in the construction of the pan closet, which renders it untrustworthy, but the one especially pointed out—the impossibility of keeping it clean—is

---

* See Chapter V., p. 107.

enough to absolutely condemn its use, from a sanitary point of view. It is decidedly the worst form of closet that can be used.

Valve closets are merely modifications of the pan closet. The bottom of the bowl is closed by a flat valve, which is held in its place by a weight. By moving a lever the valve is turned down allowing the excreta to drop into the container. The only differences between the pan and valve closets are that in the latter a flat valve is substituted for the pan of the former, and that this allows the container to be made smaller. Otherwise, there are no advantages in the valve closet. Considered from a sanitary standpoint, the valve closet is no worse than the pan closet, and but very little, if any, better.

The third variety, or plunger closet, has several marked advantages over the two just described. The characteristic feature of the closets of this class is that the outlet, which is generally on one side of the bowl, is closed by a plunger. This bowl is always from one-third to half full of water, into which the excreta fall. On raising the plunger, the entire contents of the bowl are rapidly swept out of the apparatus into the soil-pipe, the bowl thoroughly washed out by a sudden discharge of water, and on closing the outlet with the plunger, the bowl is again partly filled with water. An overflow attachment prevents accumulation of too large a quantity of water in the bowl. The Jennings, Demarest, and Hygeia, are types of this class. The principal objection is that the plunger sometimes fails to properly close the outlet, allowing the water to drain out of the bowl, and thus destroying one of its principal advantages. The mechanism is also somewhat complicated and likely to get out of order.

The hopper closet consists of a deep earthenware or enameled iron bowl, with a water-seal trap directly underneath. The excreta are received directly into the proximal end of the trap, and when the water is turned on, the sides of the bowl are washed clean and everything in the bowl and trap swept directly into the soil-pipe. There is no complicated mechanism to get out of order, the trap is always in sight and the entire apparatus can always be kept clean and inoffensive, as there are no hidden corners or angles for filth to lodge. This form of closet is, probably, all things considered, the best for general use.

The 'wash-out' closets are of various shapes, some having the trap in the bowl itself, others having a double water trap. They are generally simple in construction, and not likely to get out of order They do not present any decided advantages over the simple hopper.

Water-closets should not be enclosed in wooden casings as is almost universally done. Everything connected with the closet, soil and drain-pipes and water supply, all joints and fixtures should be exposed to view so that defects can be immediately seen and easily corrected By laying the floor and back of the closet in tiles or cement, such an arrangement can even be made ornamental, as suggested by Waring,* who says that a closet 'made of white earthenware, and standing as a white vase in a floor of white tiles, the back and side-walls being similarly tiled, there being no mechanism of any kind under the seat, is not only most cleanly and attractive in appearance, but entirely open to inspection and ventilation. The seat for this closet is simply a well-finished hard wood board, resting on cleats a little higher than the top of the vase, and hinged so that it may be conveniently turned up, exposing the closet for thorough cleansing, or for use as a urinal or slop-hopper.'

---

* Sanitary Condition of New York City. Scribner's Monthly, Vol. XXII., No. 2, June, 1881.

Where the arrangement here described is adopted, extra urinals are unnecessary, and undesirable. Where they are used they should be constantly and freely flushed with water, otherwise, they become very offensive. The floor of the urinal should be either of tiling or enameled iron.

B. Water Supply for Closets.—The water supply for flushing water-closets should not be taken directly from the common house water supply, but each closet should have an independent cistern large enough to hold a sufficient quantity of water for a thorough flushing (four to six gallons) every time the closet is used. The objections to connecting the water closet directly with the common house supply are, that there is often too little head of water to properly flush the basin, and secondly. if the water be drawn from a fixture in the lower part of the house, while the valve of a water-closet in an upper floor is open at the same time, the water will not flow in the latter (unless the supply pipe is very large), but the foul air from the closet will enter the water pipe, and may thus produce dangerous fouling of the drinking water. Hence, separate cisterns for each water-closet should always be insisted upon.

C. Traps.—Every water-closet, urinal, wash-basin, bath tub and kitchen sink should have an appropriate trap between the fixture and the soil-pipe. The trap should be placed as near the fixture as practicable; as pointed out above, in the best forms of water-closet the bottom of the closet itself forms part of the trap.

Traps differ in shape and mechanism. The simplest and most efficient are the ordinary $\bigcirc$ or half-$\bigcirc$ traps. These traps are of uniform diameter throughout and have no angles for the lodgment of filth. A free flush of water cleanses them perfectly, and they rarely fail to furnish a sufficient obstruction to the passage of sewer air through them from the soil-pipe, unless the water has evaporated, or been forced out under a back-pressure of air in the soil-pipe, or been siphoned out, and thus the seal broken.

The **D** trap and bottle trap are objectionable on account of the great liability of becoming fouled, by filth lodging in the corners, while in the mechanical traps, like Bowers' ball-valve trap, Cudell's trap and others of this class, there is always danger of insufficient seal, by filth adhering to the valve, and thus preventing its exact closure.

Most of the traps now furnished by the dealers in plumber's supplies, have an opening in the highest part for attaching a vent-pipe. It has been found that the seal in most traps can be broken by siphonage, if the pressure of air on the distal side (the side toward the soil-pipe) of the trap is diminished, or, on the other hand, by increase of pressure in the soil-pipe the water in the trap may be forced back into the fixture and thus sewer air enter the room. By providing for a free entrance and exit of air to the trap this breaking of the seal can be prevented. The ventilation of traps can however, be dispensed with, if the soil-pipe is of a proper diameter and extended through the roof. The elaborate extra system of ventilation of traps, so generally insisted upon by plumbers and sanitary engineers is unnecessary. If the soil-pipe is of the proper size and height, siphonage of traps will not be likely to occur. The waste pipe connecting the fixture and the soil pipe should be as short as possible ; in other words all water-closets, urinals, baths and lavatories should be placed as near the soil pipe as practicable, in order to have no long reaches of foul waste pipe under floors or in rooms.

D. The Soil-Pipe.—The vertical pipe connecting the water-closets and other fixtures with the house-drains, is called the soil pipe. It should be of iron, securely jointed, of an equal diameter (usually four inches) throughout, and extend from the house-drain to several feet above the highest point of the house. The connexion of all the waste-pipes from water-closets, baths, etc., should be at an acute angle, in order that an inflow at or nearly at right angles may not produce an obstruction in the free passage of air up and down the soil-pipe. The diameter of the soil-pipe at its free upper end should not be narrowed; in fact, according to Col. Geo. E. Waring, the up-draught is rendered more decided, if the upper extremity of the soil-pipe is widened.* The internal surface of the pipe should be smooth, and especial care should be taken to prevent projections inward at the joints; otherwise paper and other matters will adhere to the projections and gradually obstruct the pipe.

E. The House-Drain.—The horizontal, or slightly inclined pipe which connects the lower end of the soil-pipe with the sewer or cess-pool, the point of final discharge from the house, should be of the same diameter and material as the soil-pipe. The joints should be made with equal care, and the pipe should be exposed to view throughout while within the house-walls. If sunk below the floor of the cellar it should be laid in a covered trench, so that it may be readily inspected. The junction between the vertical and horizontal pipe should not be at a right angle, but the angle should be rounded. The drain-pipe should not be trapped. This is contrary to the advice of sanitary authorities generally, but the author thinks it unadvisable to trap the drain-pipe. There should be no obstruction to the outflow of sewage from the house, and a trap in the drain-pipe is of no avail against the passage of sewer air from the sewer or cess-pool into the soil-pipe, if

---

* Am. Architect, Sept. 15th, 1883, p. 124.

the pressure of the air in the former is increased. Furthermore, if the passage of air backward and forward between the sewer and the external air at a sufficient height (above the roofs of houses, for example), is free and unobstructed, the sewers (or the cess-pool, as the case may be) will be better ventilated than if an obstruction to such free circulation, in the form of a trap, be placed in the drain-pipe.

Nearly all sanitary authorities direct that an opening for the admission of fresh air,—'fresh air inlet,' —should be made in the drain-pipe, before its connexion with the sewer or cess-pool. This is done with the view of having a constant current of fresh air entering near the base of the soil-pipe and passing upward through it. Theoretically the current ought always to pass in this direction. Practically, however, the current is found, at times, to pass the other way, and the foul air from the soil-pipe may be discharged into the air near the ground, where it would be much more likely to do harm than when discharged high up in the air beyond the possibility of being breathed.

OFFICIAL SUPERVISION OF THE SANITARY ARRANGEMENTS OF DWELLINGS.

In most cities and towns, the municipal authorities have provided for an official inspection of buildings, to prevent neglect of precautions against fire and other manifest dangers to life. It is only very recently however that the authorities of some of the larger cities in this country have enacted laws to prevent improper construction of house drainage works. Although none of these laws or ordinances cover the subject, yet their proper enforcement must result in great advantage.

Within the past few years, following the example of Edinburg, volunteer associations have been organised in various cities of this country, with the object of securing constant expert inspection and supervision of the drainage arrangements of dwellings, and other necessary sanitary improvements.

The good results accomplished by the Newport Sanitary Protection Society, the New Orleans Auxiliary Sanitary Association and other similar bodies attests the usefulness of such organisations.

[The following works are recommended to the student who desires a fuller knowledge on the subjects treated in this chapter:

W. H. CORFIELD: Dwelling Houses, Their Sanitary Construction and Arrangements. N. Y. 1880. WM. PAUL GERHARD: House Drainage and Sanitary Plumbing; Fourth Report State Board of Health of R. I., 1881. ELIOT C. CLARKE: Common Defects in House-Drains. Tenth Annual Report Mass. State Board of Health, 1879.]

CHAPTER VII.

CONSTRUCTION OF HOSPITALS.

SITE.

IF the choice of a site for the habitations of healthy persons is a matter of vital importance, as was pointed out in the last chapter, it needs no argument to impress upon the reader the actual necessity of choosing a site with wholesome surroundings for a habitation for the sick. In selecting a site for a hospital, therefore, it is of prime importance to avoid a location where unsanitary influences prevail.

While a hospital should always be easily accessible, it is not desirable that it should be in a noisy or crowded part of a city. Where a hospital is primarily designed for the reception of accident, or 'emergency' cases, it is of course necessary to have it near to where accidents are likely to occur. In a city, this will probably be in the most crowded and noisy part.

The direction of the prevailing winds from the city should be avoided, in selecting a site for a hospital.

Free admission of sunlight and air must be secured to all parts of the hospital. An elevated location is therefore desirable, although exposure to violent winds must, if possible, be avoided.

The soil upon which a hospital is built should be clean, easily drained, with a deep ground-water level, not liable to sudden oscillations. The neighborhood of a marshy, or known malarious region should be avoided.

THE BUILDINGS.

The building area must be large enough to permit the construction of buildings in accordance with the modern recognized principles of hospital construction. Overcrowding is not permissible; either of the grounds by buildings, or of the buildings by patients.

Having determined upon the number of patients for whom provision is to be made, and the character of the diseases to be treated, an estimate must be made of the area necessary for a hospital. Taking into account all the buildings needed, the area required will be—for two or more storied buildings—not less than three hundred and twenty-five square feet per bed. If one story buildings are to be erected, more space will be required, and if infectious diseases are to be treated in the hospital, the above space allowance must be doubled or trebled. In the new Johns Hopkins Hospital in Baltimore, the area occupied by the buildings is fourteen acres, and provision is to be made for three hundred patients. This, covering, of course, the area occupied by the administration buildings, nurses' home, kitchen, dispensary, operating and autopsy theatre, laundry, etc., gives an area of 2,020 square feet per bed. The actual allowance of floor space per bed is one hundred and twenty square feet; for patients with infectious diseases, the space allowance is nearly treble, being three hundred and ten square feet.

Within recent years, the principles of hospital construction have undergone considerable modification. While formerly large hospitals consisted usually of one large, two or more storied buildings, in which all the various departments were comprised under one roof, the aim has recently been to scatter the wards as much as practicable consistent with reasonable ease of supervision and administration. Under the former plan, with large wards connected by common corridors and stairways, ease of administration was primarily secured; in the latter the most important object of a hospital, 'a place for the sick to get well in,' is more nearly attained. While many hospitals are still being constructed on the old plan, of a single block of several stories in height, nearly all sanitary authorities are agreed that the plan of separate pavilions of one, or at most two stories. in which the buildings are entirely disconnected, or connected only by means of an open corridor for convenience of administration, is best for the patients, and leaving out of account the cost of the ground, are also the most economical.

The recent development of the pavilion system of hospitals may be attributed largely to the success obtained in treating the sick and wounded in the simple barrack hospitals during the late war between the States. The army barrack hospital is the original type of the pavilion hospital of the present day.

Each pavilion consists of one or two wards, containing from ten to thirty beds altogether. In each pavilion or ward is also a bath and wash room, water-closet. dining room, scullery. attendants' room, and sometimes a day-room for patients able to be out of bed.

The two-story pavilion is built on the same plan and is generally adopted in cities. or where economy of space is desirable for financial reasons, and where no infectious diseases are treated. Where practicable, one story pavilions should always be adopted, as they are more easily heated, ventilated and served than two-storied buildings.

When a number of pavilions or wards are connected by a corridor with each other and with a central or administration building, and other service buildings, the aggregation constitutes a modern pavilion block hospital. The Johns Hopkins Hospital, already referred to, is a model hospital of this class, and its plans should be studied in detail by all who are more particularly interested in hospital construction. The general wards are in one and two story buildings, connected by a corridor with each other, and the administration and service buildings. In addition to two buildings containing private rooms and small wards for patients able to pay for the extra accommodations, there are two lines of pavilions running from east to west. The corridor cuts all the pavilions near the north ends of the buildings, separating the ward almost entirely from the service part of the building. This arrangement leaves the south, east and west fronts of the wards entirely exposed to the sun's rays, a very important advantage. The kitchen and laundry are at opposite angles of the grounds, while the autopsy building is placed in the extreme northeast corner of the grounds, as far from all the wards as practicable.

The free space between the separate pavilions should be at least twice the height of the building. In the Johns Hopkins Hospital, the space is fifty-nine feet and eight inches between the one-story common wards, which are thirty-seven feet in height from the surface of the ground to the ridge of the roof.

VENTILATION AND HEATING.

The cubic space (initial air space) per bed, in the wards should be not less than 1,500 to 2,000 cubic feet, and for surgical or lying-in cases, and contagious diseases, 2,500 cubic feet should be allowed. The ventilating arrangements should secure an entire change of the air two to three times in an hour.

In most sections of the United States, natural ventilation can be relied on to keep the air in hospital wards pure (assuming, of course, the proper construction of the buildings). The windows, doors, and walls are important factors in securing this ventilation. Hence, especial care is to be paid to their construction and arrangement.

Many German, French and English authorities on hospital building, urge the importance of making the walls impervious by cement, glass, or paint. The peculiar odor known as 'hospital odor,' it is asserted, cannot be prevented in any hospital in which the floors, walls and ceilings are not absolutely impervious. The American practice is generally in favor of walls which permit transpiration of air. In the experience of the author the imperviousness of the walls is not necessary to secure freedom from hospital odor. It remains a question for serious consideration whether the diminution of natural ventilation would not counterbalance any good resulting from non-absorptive walls.

The interior of the walls should be perfectly smooth and plain; no projections, cornices or off-sets of any kind are permissible. The desirability of this restriction was clearly expressed nearly a hundred years ago by John Howard: 'From a regard to the health of the patients, I wish to see plain, white walls in hospitals, and no article of ornamental furniture introduced.'*

Windows should run quite to the ceiling, and should not be arched, but finished square at the top. There should be one window for every two beds. The window sash should be double to retain heat, and the lights, heavy clear glass. Ventilation can be promoted by raising the outer sash from below, and lowering the inner one from above. The insertion of a Sherringham ventilator at the top of the inner sash will aid in giving the incoming air-current an upward direction.

Heating is best accomplished by introducing hot air from without, or by stoves or fire-places in the centre of the wards. Where hot air is introduced from without, it should be heated by passing it over steam

---

* An Account of the Principal Lazarettos of Europe, etc. London, 1791, p. 57.

or hot water coils, and not by passing it through a furnace, which may produce super-heating and excessive dryness of the air.

In series of experiments by Dr. Edward Cowles at the Boston City Hospital,* the air was heated to 90° by passing it over steam coils. It was admitted to the wards by numerous inlets, one foot square. The best velocity for ventilating and warming purposes was found to be 180 feet per minute. Exit openings were in the ceiling, and it was found best to make them large, as by this means the rapidity of exit currents is reduced.

Where the warming of the ward must be accomplished by stoves or fire-places in the ward, the best plan, for square and octagon wards, is to have a large central chimney with arrangements on the four sides for fire-places or stoves. This chimney can also be used as a very efficient ventilating shaft throughout the year, by a device put in practice by Mr. John R. Neirnsee, architect of the Johns Hopkins Hospital.† In oblong wards, two or more large stoves placed at equal distances along the centre of the wards will heat the wards effectually.

Floors should be made of tiles, slate, or oak or yellow pine lumber. If wood is used, it should be well seasoned, perfectly smooth, and all joints accurately made. The floor should be kept constantly waxed to render it impervious to fluids.

The space between the floor and ceiling below should be filled with some fire-proof non-conducting material, such as cement, or hollow bricks, in order to isolate each floor or ward as much as possible from others, both to prevent transmission of noise and extension of fire.

---

* Report of the Massachusetts State Board of Health for 1879, p, 231-248.

† Hospital Construction and Organisation. Plans for Johns Hopkins Hospital, N. Y., 1875, p. 325 ,et. seq.

All corners and angles on the inside of the building should be rounded to facilitate the removal of dust.

In cleaning up, care should be taken not to stir up the layers of dust too much, by active sweeping or dusting. The floors, furniture, door and window-casings should be wiped off with damp cloths. Soiled bedding, clothing, dressings and bandages must be promptly removed from the ward. Mattresses and other bed-clothing should not be shaken in the ward.*

Water-closets—or where the dry method of removal of excreta is in use—earth or pail-closets should be placed where they can be easily reached by the patients. but the apartment in which they are placed must not open directly into the ward. The entrance to this apartment should be from the corridor, or better still, from the open air. The ventilation of water-closets should be independent of, and entirely distinct from that of the ward, or other part of the hospital building.

It is, of course, unnecessary to more than call attention to the vital importance of the prompt removal of all excreta, both solid and liquid, from the ward or hospital building. To attempt disinfection of excreta and allow them to remain in the ward after being voided, is a pernicious practice, which should under no circumstances be permitted. All utensils for the reception of excreta, bed pans, etc., should be immediately emptied and thoroughly cleansed.

Urinals are not advisable; the simple hopper closet with hinged hard-wood seat, as described in chapter VI. is sufficient.

A bath room and lavatory should be attached to every ward. It should be placed in the service build-

---

* A Wernich: Ueber Verdorbene Luft in Krankenræumen. Volkmann's Samml. Klin, Vortr., No. 179, p. 24.

ing, and be easily accessible to the patients. There should also be portable bath tubs in order that baths may be given in the wards when necessary.

Every large general hospital should also have a special apartment or building where baths of various kinds, such as medicated, vapor, Turkish and Russian baths could be given. In lying-in hospitals, special arrangements for giving vaginal and uterine douches must also be furnished.

A daily water-supply of at least one hundred gallons per bed should be provided. The water should be easily accessible from the wards and various parts of the service building.

All water-closets, soil and waste-pipes must be properly trapped; all joints must be properly made and all sewer connexions made on the most improved plans. All work of this sort should be properly tested before being accepted, and frequently inspected afterward.

No sewer or house-drain should be laid under a ward.

A disinfecting chest for disinfecting soiled clothing, bedding, dressings, etc., should be placed in the basement of the ward, and connected with the latter by an iron chute, closing perfectly by an iron top. The best and most convenient disinfectant is steam. This is also the best means to destroy vermin in clothing and bedding.

It is questionable whether a nurses' room should be attached to a hospital ward. The nurse's place, when on duty, is in the ward itself, not in a room separate from it. Where there is a nurses' room, it should not be furnished with sleeping arrangements, for this is a strong temptation to neglect of duty on the part of the nurse. A nurse not on duty should not be permitted to remain about the ward.

A ward-kitchen should be in the service building, where articles of food can be kept hot, or cold, when necessary, and where special dressings, cataplasms, hot water, etc., can be prepared. A small gas-stove only should be allowed in the ward-kitchen, as the regular meals of the patients are prepared in the central kitchen, which should be totally detached from the hospital. The ward-kitchen can be easily utilised as a nurses' room, and in a small hospital can also be used as a store-room for the patients' body and bed linen, and clothing.

The dining room for patients able to be out of bed should be in the service building. A room with good light, and well ventilated and heated should be selected for this purpose. In the intervals between meals this room could be used as a day room for such patients as should be out of bed, but who are not able to be in the open air.

A dead-house, containing a dead room, autopsy room, and a room fitted up for rough microscopic, and possibly, photographic work, is a necessity to every well-appointed general hospital. The dead-house should be entirely separate from the ward buildings.

The kitchen should be separate from the other buildings, and, in large hospitals should also be the central station for the heating arrangements, if hot water or steam heat are used. The laundry may be connected with it. The kitchen should be connected with the wards by means of a covered corridor to avoid exposure in carrying the food to the wards.

The administration building should contain office rooms for the superintendent and resident physician, pharmacy, library, reception rooms for visitors, living rooms for one or more assistants, and dwellings for the superintendent and resident physician.

THE ADMINISTRATION AND MANAGEMENT OF A GENERAL HOSPITAL.

The general management of a hospital should be under the direction of a superintendent, who besides being a medical man, should be especially qualified by study and experience for the work. The superintendent of a large hospital should not be expected to perform any of the routine professional work in the wards, but he should be responsible for the service, both professional and lay, in the hospital. He should be the financial officer, and in all things concerning the hospital his judgment should decide. He should have sufficient assistance to permit all necessary duties to be promptly performed. For this purpose, he should have a secretary, or clerk, who should not be a medical man; otherwise the attention of the latter might be withdrawn from his clerical duties to the more interesting professional work in the hospital. The plan advocated by some authorities, to have two superintendents for large hospitals, one of whom shall be a medical man and direct only the professional work of the hospital, while the other shall have charge of the administrative functions, does not commend itself to the author. It involves a division of responsibility which will in nearly all cases, eventually lead to differences of opinion likely to prove unfavorable to the best interests of the hospital.

It is customary in this country to appoint as resident physicians and surgeons in hospitals, recent graduates, whose functions are usually limited to carrying out the directions of the visiting physicians and surgeons, and sometimes to act on their own responsibility in emergencies. This system has some advantages for the physicians, but is usually detrimental to the best interests of the patients. The resident medical officer in a large hospital should always be a thoroughly qualified, experienced physician, capable of deciding promptly when the occasion arises, and he should be responsible to the superintendent for the proper performance of his professional duties. Necessarily, a physician with the qualifications indicated, would demand a very much larger salary than is usually paid resident physicians, but it should be understood, that no hospital, in which the good of the patient is the first consideration can be conducted on a cheap basis.

Visiting physicians and surgeons, and all resident medical officers should be chosen with reference to their general and special qualifications for the duties expected of them. It would seem to be a good plan to make the selections for subordinate positions at least, by competitive examination.

The sick in a hospital should be properly classified. Male and female patients, should of course be treated in separate wards. A primary classification into medical, surgical and obstetrical cases or wards, is also indicated. Infectious diseases, such as typhoid fever, erysipelas, cholera, yellow fever, croupous pneumonia, etc., should not be treated in the same wards with rheumatism, Bright's disease, cardiac and nervous disorders, or simple digestive derangements. It is questionable, however, whether it is advisable to make a very elaborate classification of the various diseases, except in very large hospitals.

An accurate record, made at the time of observation, and not written from memory afterward, should be kept, of the history and progress of every case. The record should show, not merely the symptoms and diagnosis, but the medical and hygienic treatment. In most hospitals where such records are kept, the entries are made either in a simple memorandum book, or in a more or less complicated case record. A simple form of case record has been devised by Surgeon Walter

Wyman, U. S. Marine Hospital Service, which seems to possess advantages that render its general adoption desirable.

In hospitals where cases of surgical diseases and injuries are received, a special apartment should be fitted up as an operating room. Operations should not be performed in a ward in the presence of other patients.

[The following works are recommended for additional study upon this subject:

Hospital Construction and Organisation; N. Y. 1875. (especially the essays of Drs. BILLINGS, FOLSOM and STEPHEN SMITH.) Krankenanstalten, by L. DEGEN, in V. PETTENKOFER UND ZIEMSSEN's Handbuch der Hygiene. Spital, by C. BÖHM, in Realencyclopädie d. ges. Heilkunde. Bd. XII. General Principles of Hospital Construction, by F. H. BROWN, in BUCK's Hygiene and Public Health, Vol. I.]

CHAPTER VIII.

SCHOOLS.

THE hygiene of schools comprises the consideration of the sanitary principles underlying the construction of school-houses and school furniture, the proper amount of time to be devoted to study at different ages, the special diseases of school-children, their causes, and means for their prevention.

CONSTRUCTION OF SCHOOL HOUSES.

In the construction of school-houses, the same hygienic principles are applicable as in dwelling-house construction. The selection of a site for the school-building should command the same careful consideration that is necessary in determining upon a site for a dwelling. Proximity to marshes and other unsanitary surroundings should be avoided. If the soil is damp, it should be properly drained, and all sources of insalubrity in the neighborhood avoided, or, if possible, removed.

School houses should not be over three stories high; corridors and stairways should be wide, straight, and well lighted. All stairs should be securely built, and be guarded with ample, strong railing. All doors should open outward to permit ready egress, and reduce the danger of accident in panics from any cause.

In addition to the study, or recitation rooms, provision should be made for play and calisthenic exercise rooms. Well-lighted and ventilated side-rooms should be provided for the reception of outside clothing, umbrellas, overshoes, etc. These articles should not be kept in the recitation or study rooms.

Floors should be made of accurately jointed flooring, rendered impervious by oil or paraffin coating.

Appropriate measures must be employed to prevent the permeation of the building by ground-air.

The inside walls of school-rooms may be tinted a neutral gray, or light blue or green. Ceilings should be white. Walls and ceilings should not be painted, but lime-coated to permit free transpiration of air.

Schools should be so constructed as to permit of ready heating, and ventilation, cleaning and keeping clean. In large schools, the method of heating will usually be by furnace-heated air, although a better method would probably be by steam or hot-water pipes.

The ventilation of school-rooms must be carried out on the principles indicated in chapter I. With careful and intelligent teachers, natural ventilation will give better satisfaction than a complicated artificial system.

A model study room, according to modern views, should be about thirty feet long, not over twenty-two feet wide and fourteen feet high. Such a room could be easily lighted by windows on one side only, and readily heated and ventilated. It would also enable the teacher to exercise a close supervision over his pupils. In a room of this size forty pupils would be a proper number, although fifty could be accommodated. The initial air-space for each pupil would be 135 cubic feet if there were fifty pupils in the room, and 231 cubic feet if there were only forty. This would be slightly reduced by the allowance for the teacher.

It is believed that study rooms should face toward the north. The light entering from the north side of a building would be equable during the whole day. While a larger window surface would be necessary than with an easterly or southerly exposure, it is held that the light being devoid of all glare, would be more effective. Where the light is admitted on the east, south, or west sides of the building, the direct entrance of the sun's rays must be prevented by curtains, by means of which the amount and proper distribution of the light is regulated with difficulty.

The windows of the school-room should reach from about the height of the pupils' shoulders (when seated) to nearly, or quite to the ceiling. Arches or overhanging cornices over the windows should be avoided, as they cut off much light. For the same reason the near proximity of other high buildings, and of trees should be avoided in selecting a site for a school-house. The window area should be not less than one-fifth of the floor area, otherwise the light will be deficient.

The light should be admitted only from the left side of the pupil. When admitted from the right side the shadow cast by the pen in writing interferes with good vision; if admitted directly in front of the pupil the glare of the light will injuriously affect the eyes, while if it enters from behind, the book or paper of the pupil will be so much in shadow as to compel him to lean so far to the front in bringing his eyes nearer to book or paper, that nearsightedness is very likely to be developed. Furthermore, if the light is admitted into the room at the backs of the pupils, the eyes of the teacher are liable to suffer from the constant glare.

In a school-room of the dimensions above stated, a row of windows on one side, forming an area of glass one-fifth of the floor-space, will thoroughly and satisfactorily illuminate the room, with the least unfavorable influence upon the organs of vision. It is advisable, therefore, to always insist on this arrangement of lighting of school rooms. Where artificial light is used in a school-room, it should be in the proportion of one burner to every four pupils. All burners should be provided with chimneys and vertical reflectors.

Water-closets and privies should not be placed in cellars or basements. This would seem to be self-evident, and yet, in many city school-houses these places of retirement are in this unsuitable location. When it is considered that large schools are frequently warmed by hot air taken from the cellar, it furnishes an additional reason to avoid this location for water-closets. On the contrary, the custom, in some country schools, of placing the privy at a considerable distance from the school-room and in an exposed situation, is almost equally reprehensible, as the pupils, especially girls, are prone to neglect obeying the calls of nature, from which neglect many disorders arise.

SCHOOL FURNITURE.

Desks should be slightly sloping, the edge nearest the pupil being about one inch higher than his elbows. The front edge of the seat should project a little beyond the near edge of the desk, so that a plumb-line dropped from the latter should strike the seat near its front edge. If the seat is not thus brought slightly under the desk, the pupil is compelled to lean forward in writing, which position prevents proper expansion of the chest and increases the blood-pressure in the eyes, a condition promotive of nearsightedness.

Seats should be only high enough so that the feet rest flat upon the floor. If they are higher, a foot-board must be provided. Children should not be condemned to the cruelty of having their feet dangling 'between heaven and earth' while they keep their seats. Seats and desks should be graded according to the sizes of the pupils—not their ages or standing in the class.

An ideal seat and desk would be one made to measure for each pupil, but this is manifestly impracticable, inasmuch as with the constant growth of the child the seats would be rapidly outgrown.

Blackboards should not be placed at a greater distance than thirty feet from the farthest pupil. The ground of the board should be a dead black, without lustre. In writing exercises upon the board, care should be taken that the letters and figures are made sufficiently large, and with rather heavy strokes of the crayon, in order that they may be easily seen from the most distant part of the room. It has recently been demonstrated that a black letter on a white ground can be seen at a greater distance than a white letter on a black ground. Hence, it might prove advantageous to the eye-sight of school-children, to substitute for the present blackboard and chalk, a white board and black crayon. In some European lecture rooms, this plan has been adopted with satisfaction.

AMOUNT OF TIME TO BE DEVOTED TO STUDY.

Young children should not be kept at the same study, or in the same position for long at a time. The exercises should be frequently varied. It is especially with children in the primary grades that care should be taken not to overburden their minds with too many hours of study, or too long continuance at the same exercise.

Children should not be placed in school much, if at all, before the completion of their seventh year. From seven to nine years, they should be kept at their studies not longer than three hours daily; from nine to twelve years, four hours may be allotted them, and from twelve to sixteen years, they may be kept at mental work five to six hours daily. This does not mean that pupils are to be kept continuously at their studies during these hours, but that they should be neither compelled, nor permitted to study longer than these periods each day. It is believed that these figures represent the capacity for endurance in the majority of children, and they should be adopted in all schools where the largest return in mental acquirements is desired at the least expenditure of health. Excess of time expended in study is almost certainly followed by physical deterioration.

Gymnastic exercises should form part of the daily routine in all schools. These exercises should take place, when practicable, in the open air. Playing, romping, laughing and singing should be encouraged, rather than the natural tendency to boisterous play restrained. It is especially desirable that the female children should be encouraged to take part in these diversions. The desire, on the part of many parents, to see little girls deport themselves as young ladies, before the time even when they write their age in two figures, is very reprehensible, and deserves the most unqualified condemnation.

DISEASES OF SCHOOL CHILDREN.

The principal diseases incident to school life are, myopia, spinal deformities, nervous and digestive disorders, pulmonary phthisis, and contagious diseases. It is believed that by judicious sanitary measures these can all be very much diminished and some entirely prevented.

It has been shown by the examination of the eyes of school-children, that nearsightedness increases progressively from the lowest to the highest classes. Children who enter school with an hereditary tendency to myopia or who are perhaps already nearsighted to a slight degree, soon become more intensely myopic, while others who may be even hypermetropic on entering school will be found to have become nearsighted during school life. In examinations of over 30,000 pupils of grammar and high schools, in Germany, Austria, Russia, and Switzerland, it has been found that the average proportion of nearsightedness is a fraction over 40 per cent., varying in the different classes from 22 per cent. for the lowest, to 58 per cent. for the highest classes. These figures represent the averages of all the examinations made. In some particular schools, for example, in the gymnasium (high school) of Erlangen, the percentage in the higher classes was 88 per cent., in the gymnasium of Coburg, 86 per cent., and in the gymnasium of Heidelberg the proportion of myopic students in the highest class is said to have reached 100 per cent. in 1877. In the primary schools the percentage was found to be much lower. These observations show that the number of myopic individuals bears a constant relation to the intensity of use of the visual organs. The results of the observations of different observers in different countries also uniformly point to the conclusion that not only does the number of nearsighted pupils increase as the higher classes are reached, but the degree of myopia increases likewise. Thus a pupil who may have only a moderate degree of myopia on entering the school will have myopia in a higher degree as he advances in his classes. Erismann found, on re-examining the same pupils annually, that in six years, 13.14 per cent. of those examined had developed myopia from emmetropia, while in 24.57 per cent. of nearsighted pupils, the degree of myopia had increased.[*]

The principal causes of the prevalence of near-sightedness in schools are badly arranged, or insufficient light, bad air, overheating of the school-rooms, improper construction of desks, compelling children to lean forward while reading or writing, and badly printed text-books. The use of small type, poor paper, and bad press-work in text-books is very reprehensible.

---

[*] ERISMANN: Die Hygiene der Schule, in VON PETTENKOFER und ZIEMSSEN's Hand buch der Hygiene. II Th., 2 Abth., p. 30.

The type technically known as *long primer*, one size smaller than that used in this book, is the smallest that should be used in text-books. That badly arranged light and improper seats are causes of myopia, has been shown by Florschutz in his examinations of the pupils in the public schools of Coburg. He found that in the newer schools in which the light and seats are better arranged, the percentage of nearsight decreased. The average percentage of those examined in 1874 was twenty-one, while in 1877 it had been reduced to fifteen,* showing the great improvement due to the application of correct sanitary principles in construction of school-houses.

Defective hearing has recently been shown to be especially frequent among school children. A Berlin aurist found 1,392 children out of 5,902 (23.6 per cent.) suffering from ear disease of some kind. Dr. Samuel Sexton, of New York, and Dr. Chas. F. Percivall, director of music in the public schools of Baltimore, have arrived at similar results after examination of a large number of school children.

Spinal curvature is present in a large proportion of the children attending schools. Statistics are not very full upon this subject, but one author, Guillaume, states that he found lateral curvature of the spine in 218 out of 731 school children, a proportion of 29.5 per cent. This, of course, includes the slighter degrees of curvature, which cannot properly be termed a disease. M. Eulenburg† found that among 1,000 persons with lateral curvature of the spine, the disease began in 887 between the ages of six and fourteen, that is to say, during the years of school life. Girls are affected ten times as often as boys.

The especial causes of spinal curvature occurring during school life, are improperly constructed seats

---

* Quoted by Cohn in Realencyclopædie d. ges. Heilk. Bd. 12, p. 263.
† Realencyclopædie d. ges. Heilk. Bd. XI. p. 564.

and desks, and an improper position of the body. Many pupils habitually assume a 'twisted' position, which is very liable to produce spinal distortion in children of weak muscular development. An improper position is more likely to be unconsciously assumed by girls than by boys. The clothing is responsible for this, for when the girl files into her place behind the desk, her clothing, hanging loosely about her, is swept back, and forms a pad, upon which she sits with one buttock. The greater elevation of her seat on that side throws the spinal column out of the vertical line, which is compensated by a partial twisting of the trunk. The attention of teachers should be directed to this faulty habit, which can be easily corrected, and its consequences averted by timely interference.

Nervous disorders are comparatively frequent among school-children. Headaches are often due to insufficient ventilation, improper food, bad digestion, and excessive mental strain. Defective light may also be the cause of headaches, by causing ocular fatigue. Hysterical and imitative affections are not infrequent, and sometimes pass through entire schools, including even the teachers. Girls are of course more subject to this class of disorders than boys, but the latter are not entirely exempt.

Derangements of the digestive organs are exceedingly frequent among school-children. They can generally be traced to the use of improper food. The eating of cold lunches should be discouraged as much as possible.

Nuts, candies, pies, fruit-cakes, and above all pickles, are most fruitful sources of digestive derangements of children. The absence of proper accommodations to enable children—especially girls—to answer the demands of nature, are frequent sources of digestive and nervous disorders.

The seeds of pulmonary consumption are frequently implanted during school-life. A neglected cough, bad ventilation, under which term may be comprised over-heating and cold draughts, as well as polluted air, improper position of the body, excessive mental work, or underfeeding, may, any of them be the starting point of this fatal disease.

Especial care should be taken to prevent the introduction or dissemination of contagious diseases through schools. The importance of this duty should be at all times impressed upon school boards and teachers. In the first place, no child should be admitted within the door of the school-room, unless it first presents undoubted evidence of protection againt small-pox, either by having passed through a previous attack, or by a proper vaccination. In case of an actual or threatened epidemic of small-pox, the entire school, including teachers, should be vaccinated.

Diphtheria has been shown to be frequently spread through the agency of schools.* This fatal disease demands especial precautions on the part of teachers, and others involved in the management of schools, to prevent its introduction to these institutions.

Children should not be admitted to school, coming from a house where there is at the time, or has recently been, a case of contagious disease, such as small-pox, diphtheria, scarlet fever, or measles. At least four weeks should be allowed to elapse after the termination of such disease before a child from the infected house is re-admitted to the school. It goes without saying that no child having itself been sick with a contagious disease should be admitted to school until entirely restored to health. The above-mentioned limit of four weeks

---

* H. B. BAKER. The Relations of Schools to Diphtheria and to Similar Diseases. Public Health, Vol. VI, p. 107.

is the briefest period allowable before the quarantine of the infected house (so far as the schools are concerned) can be relaxed.

When a case of contagious disease has accidentally obtained entrance to the school, the pupils should be dismissed for the day, and the room thoroughly disinfected by means of sulphur or chlorine.

Teachers are not infrequently guilty of the grave imprudence of sending pupils from the school to the house of an absent child to inquire the reason of the latter's non-appearance at school. It frequently happens that the absent child is sick, and the messenger is invited to the sick-room to see his or her class-mate. There can be no room for doubt that scarlet fever, diphtheria and measles have often been introduced into schools in consequence of such thoughtlessness on the part of teachers.

In order to promote the proper hygienic management of schools, all teachers should be required to submit to an examination in the principles and practice of hygiene, at least so far as school hygiene especially is concerned. This is a demand that school-boards could reasonably insist upon, and there can be no question that the improvement in the health of the pupils would amply justify the innovation.

[Students may consult with advantage the following special articles:

D. F. LINCOLN: School Hygiene, in BUCK's Hygiene and Public Health, Vol. II. F. ERISMANN: Die Hygiene der Schule, in VON PETTENKOFER UND ZIEMSSEN's Handb. d. Hygiene, II, Th. 2 Abth. REUSS: Schulbankfrage, in Realencyclopädie d. ges. Heilk., Bd. XII. H. COHN: Schulkinderaugen, ibid. C. J. LUNDY: School Hygiene. Public Health, Vol. IX.

CHAPTER IX.

INDUSTRIAL HYGIENE.

ONE of the most interesting chapters in the study of hygiene is that which treats of the relations of occupations to health and life. While it is unquestionable that certain occupations are intrinsically dangerous to health, there can be no doubt that in many instances incidental conditions not necessarily connected with the occupation are factors in the production of diseases. Such factors are bad ventilation and other insanitary surroundings, as well as in many cases, want of sufficient or proper food.

Occupations induce disease by compelling the workmen to inhale irritating, poisonous, or offensive gases, vapors, or dust; or by causing the absorption through the skin or mucous membranes, of irritating or poisonous substances. Changes of temperature, as exposure to great heat or cold, produce diseases which are, in some instances, characteristic. In another class of cases the excessive use of certain organs, as the nervous system, the eyes, the vocal organs, or various groups of muscles produce characteristic morbid effects. Again, a constrained attitude while at work, a sedentary life, or occupations involving exposure to mechanical violence are recognised sources of disease and death.

The following table gives the mortality and average age at death of all decedents over twenty years of age, whose occupation was specified, in the State of Massachusetts, for thirty-one years and eight months. The total number of decedents was 144,954; the average age at death, 50.90 years. Subdivided into classes, and individual occupations, the results are as follows:

TABLE I.

*Occupations of Persons whose Occupations were specified, and whose Deaths were registered in Massachusetts during a period of thirty-one years and eight months, ending with December 31, 1874.**

OCCUPATIONS.	NUMBER OF PERSONS.	AVERAGE AGE AT DEATH.	OCCUPATIONS.	NUMBER OF PERSONS.	AVERAGE AGE AT DEATH.
CLASS I. *Cultivators of the Earth:* Farmers, Gardeners, etc.	31,832	65.29	Paper-makers	288	48 29
CLASS II. *Active Mechanics abroad:*	10,893	56.19	Piano-forte-makers	111	43.33
Brickmakers	106	46.85	Plumbers	131	35 53
Carpenters and Joiners	6,150	53.33	Potters	40	56 67
Caulkers and Gravers	180	58.59	Pump and Block-makers	89	54.79
Masons	1,662	50.33	Reed-makers	9	42 78
Millwrights	118	59.14	Rope-makers	248	53.05
Riggers	161	52.25	Tallow-chandlers	67	54.93
Ship-Carpenters	873	58.53	Tin-smiths	375	41 05
Slaters	81	40.99	Trunk-makers	48	39.60
Stone-cutters	1,025	40 90	Upholsterers	124	38 82
Tanners	537	50.36	Weavers	480	44.95
CLASS III. *Active Mechanics in shops:*	16,576	47 57	Wheelwrights	507	56.98
Bakers	471	47.04	Wood-turners	76	52.07
Blacksmiths	2,402	53 26	Mechanics (not specified)	2,015	44.84
Brewers	28	47.11	CLASS IV. *Inactive Mechanics in shops:*	17,233	43 87
Cabinet-Makers	781	48.84	Barbers	403	39 81
Calico-printers	9	52.11	Basket-makers	70	61.63
Card-makers	39	48.23	Book-binders	150	40 12
Carriage-makers and Trimmers	276	48 21	Brush-makers	53	43 11
Chair-makers	138	41.77	Carvers	90	34.00
Clothiers	84	56.50	Cigar-makers	154	3*.36
Confectioners	85	44.11	Clock and watch-makers	100	52 86
Cooks	112	40.82	Comb-makers	134	51.33
Coopers	927	59.22	Engravers	124	40.88
Coppersmiths	101	45.89	Glass-cutters	76	43.16
Curriers	366	41.50	Harness-makers	423	48.74
Cutlers	131	39.21	Jewellers	468	40.34
Distillers	27	56.85	Operatives	2,138	39.16
Dyers	143	45.17	Printers	717	38.62
Founders	361	42.51	Sail-makers	217	53 21
Furnace-men	133	43.42	Shoe-cutters	362	42.34
Glass-blowers	132	37 88	Shoe-makers	9,772	44.61
Gun-smiths	250	48.86	Silversmiths or Goldsmiths	92	46.13
Hatters	356	54.67	Tailors	1,393	47.34
Leather-dressers	179	47.23	Tobacconists	43	50 35
Machinists	2,097	41 67	Whip-makers	99	43.63
Millers	278	57.14	Wool-sorters	155	48.09
Musical Instrument-makers	33	46.73	CLASS V. *Laborers* (no special trades).	28,058	47 41
Nail-makers	174	41.49	Laborers	27,382	47.49
Pail and Tub-makers	5	36.60	Servants	389	40 10
Painters	1,850	45.07	Stevedores	76	52 09
			Watchmen	193	50.06

* Thirty-third Registration Report of Massachusetts, p. CVII., et. seq.

TABLE I. *Continued.*

OCCUPATIONS.	NUMBER OF PERSONS.	AVERAGE AGE AT DEATH.	OCCUPATIONS.	NUMBER OF PERSONS.	AVERAGE AGE AT DEATH.
Workmen in Powder-mills.............	18	39.67	Manufacturers.......	1,375	51.53
CLASS VI. *Factors laboring abroad, etc.*:.	7,035	36.29	Merchants..........	3,927	54.17
			Newsdealers and Carriers	27	41.22
Baggage-masters.....	37	34.08	R. R. Agents or Conductors..........	318	39.85
Brakesmen	246	26.54			
Butchers	537	50.19	Saloon and Restaurant keepers......	299	40.90
Chimney-sweeps	4	34.50			
Drivers.............	327	38.88	Stove Dealers......	12	45.25
Drovers.....	17	49.29	Telegraphers........	5	28.80
Engineers and Firemen........... ..	567	38.77	Traders...........	2,908	48.08
Expressmen........	216	41.30	CLASS IX. *Professional men*:..	5,175	50 81
Ferrymen...........	9	53 78	Architects..........	29	47.07
Lighthouse-keepers..	10	60 40	Artists	186	44.18
Peddlers...........	417	45.18	Civil Engineers......	117	42.32
Sextons.............	81	59.94	Clergymen.	965	58.57
Soldiers............	2,885	28.37	Comedians.........	32	37 31
Stablers............	354	42.54	Dentists............	114	41.61
Teamsters..........	1,282	40 35	Editors and Rep'trs..	87	46.68
Weighers and Gaug'rs.	24	60.67	Judges and Justices..	18	64.11
Wharfingers.	22	50.00	Lawyers..........	676	56.45
CLASS VII. *Employed on the Ocean*	8,844	46.44	Musicians	266	41.59
Fishermen..........	433	42.82	Photographers......	10	36.80
Marines	4	41.25	Physicians..........	1,166	54.99
Naval Officers.......	58	50 00	Professors..........	45	55.93
Pilots..............	82	60 38	Public Officers.....	437	55.37
Seamen	8,267	46.45	Sheriffs, Constables & Policemen........	158	53.76
CLASS VIII. *Merchants, Financ'rs, Ag'ts,etc.*:	15,977	48 95	Students............	288	23.93
Agents.............	376	46.76	Surveyors...........	86	51.44
Bankers............	49	57.61	Teachers	495	41.79
Bank Officers..	151	55.14	CLASS X. *Females*:..	3,343	39.13
Boarding House k'prs.	75	47 96	Domestics	1,037	46.64
Booksellers.........	73	53 05	Dressmakers........	259	43.36
Brokers............	198	49.58	Milliners...........	186	39.42
Clerks and Bookkeepers...............	3,435	35.98	Nurses.	116	61.06
			Operatives..........	708	27 82
Druggists and Apothecaries..........	255	42 37	Seamstresses	289	46.50
			Shoe-binders........	48	43.12
Gentlemen..........	1,512	68.42	Straw-workers	73	34.83
Grocers............	517	47.59	Tailoresses.........	233	47.49
Innkeepers	467	50.04	Teachers............	442	31.27
			Telegraphers ...	7	24 48

The above table cannot be absolutely relied upon
for several reasons, the principal of which is that the
table is incomplete. It shows, however, very clearly
the relations of certain occupations to longevity. It
is seen, for example, that agriculturists have the greatest expectation of life. Next to these come mechanics

engaged out of doors. Professional men come next, and of these, clergymen and members of the bar have the first and second places, respectively. The expectation of life of physicians is above the average, being nearly fifty-five years. Mechanics engaged in active work indoors, may expect to live 3.70 years longer than those whose occupation requires them to retain a more or less constant position.

Occupations which are accompanied by the formation of much dust, either inorganic or organic, are especially unfavorable. They usually produce diseases of the respiratory organs, which may eventuate in phthisis. In the table it is seen that the average age at death of stone-cutters was 40.90; of cotton factory operatives (male), 39.16; (female), 27.82;* of cigar makers, 38.36, and of cutlers, 39.21 years. These figures more or less closely approximate the conditions which have been shown to exist in England and on the continent of Europe. In Sheffield, the workmen who grind and polish the cutlery, called 'dry grinders,' are said to suffer from a characteristic pulmonary affection termed 'grinder's asthma,' (emphysema) in the proportion of 69 per cent. of the whole number employed. The average duration of life of the needle grinders of Derbyshire is 30.66 years. Among the cutlery grinders of Solingen in Rhenish Prussia, Oldendorff found 29 per cent. suffering from pulmonary affections, while the average age at death of the 'dry grinders' was 40.7 years.

OCCUPATIONS PREJUDICIAL TO HEALTH.

The diseases of occupations may conveniently be divided into the following classes:

---

* These figures must be accepted with much reserve. While it is probable that the average age at death among women engaged in different occupations is less than that of men engaged in the same occupations, the figures in table I., class X. cannot be used as a basis of comparison. So many women are annually withdrawn from the various occupations, by marriage, which places them under different conditions that the statistics of the occupations of women in the table are untrustworthy.

1. Diseases due to the inhalation of irritating or poisonous gases and vapors.

2. Diseases due to the inhalation of irritating or poisonous dust.

3. Diseases due to the absorption or local action of irritating or poisonous substances.

4. Diseases due to exposure to elevated or variable temperature or atmospheric pressure.

5. Diseases due to excessive use of certain organs.

6. Diseases due to a constrained attitude and sedentary life.

7. Diseases from exposure to mechanical violence.

I.—DISEASES DUE TO THE INHALATION OF IRRITATING OR POISONOUS GASES OR VAPORS.

Sulphuric acid gas is used in various trades as a bleaching agent. In the manufacture of straw hats, and in the drying or 'processing' of hops this agent is extensively employed, and the people engaged in these industries frequently suffer from respiratory and digestive disorders. These are, however, rarely serious. If free access of air is allowed, the dangers to health in the above employments are very slight.

Nitric acid fumes may be dangerous to health when inhaled in a concentrated form, but very few cases are on record where any positively deleterious influence can be traced to this agent.

Hydrochloric acid fumes may prove deleterious to the workmen in soda manufactories, where the fumes are disengaged during the so-called 'sulphate process.' But the danger is probably slight. On the other hand, attention has recently been called to a peculiar effect of hydrochloric acid fumes upon the workmen in fruit-canning establishments. The men who seal or 'cap' the cans after being filled, are the ones affected. The lesion has been described by Dr. W. Stump For-

wood, who says concerning it. 'The constant inhalation
of the fumes of muriatic acid, associated as they are
with the lead solder, which the busy "capper" neglects
to protect himself against, soon produces inflammation
of the mucous membrane of the nose, which finally
results in ulceration. With some patients, after the
removal of the cause, and the application of proper
treatment, recovery takes place after two or three
months ; but with those who have a scrofulous taint
in their constitutions, this ulceration is exceedingly
intractable, and in spite of all treatment proceeds for
months and even years, until the septum is finally perforated. And strange to say, it is the common experience of those who have suffered, that as soon as perforation takes place, all the soreness and consequent
annoyance disappears and the patient recovers, with,
of course, a permanent opening in the nasal septum.'*
Dr. Forwood adds that anointing the nose both within
and without several times a day, and avoidance of the
acid fumes as much as possible, will prevent the peculiar affection described.

Ammonia has rarely caused disturbances of health
in workmen brought into contact with it. When
present in the air in large proportion it may give rise
to serious symptoms. As it is often used to prevent
the poisonous effects of mercury (q. v.) care should be
taken that the proportion of the vapor in the air of
the workroom should not exceed five per cent.

Chlorine gas is very deleterious in its effects upon
the workmen brought in contact with it in the various
industries in which it is employed. Nearly one-half
of the workmen engaged in the manufacture of chlori-
nated lime, and in bleaching, become affected.

The respiratory organs are principally affected.
Pneumonia is exceptionally frequent. If an affected

---

* Phila. Med. and Surg. Reporter, June 30, 1883.

individual is predisposed to consumption the latter disease is soon lighted up, and proves quickly fatal. The
effect of the inhalation of concentrated chlorine is thus
graphically described by Hirt.* 'The workman suffers
from violent cough, and extreme dyspnea. In spite of
the aid of the auxiliary respiratory muscles, the entrance
of air to the lungs is insufficient, and the widely opened
eyes, the pale bluish color, and the cold perspiration
plainly show the mortal agony of the patient. With
this the pulse is small, the temperature decreased.
Soon after removal from the impregnated atmosphere,
these phenomena disappear, and a few hours later, the
workman is found enveloped in chlorine and hydrochloric acid vapors in his accustomed place in the
factory. The attacks seem to be but rarely fatal.'

The constant inhalation of an atmosphere strongly
impregnated with chlorine produces a cachectic appearance, bronchial catarrh, loss of the sense of smell, and
a prematurely aged appearance. When this stage of
chronic chlorine poisoning has been reached, complete
health can rarely be re-established, even if the patients
be entirely removed from the irritating atmosphere.

Carbonic oxide is often present in the air of gas
works, iron smelting works, and coke, or charcoal
furnaces. The workmen engaged in these industries
often suffer with diseases of the respiratory organs,
digestive disturbances and general debility. Acute
poisoning from carbonic oxide is relatively frequent, as
already pointed out.† The prominent symptoms are at
first violent headache, dizziness and roaring in the ears.
These symptoms are followed by great depression of
muscular power, nausea and vomiting. The vomited
matters sometimes gain entrance into the trachea, and
may thus produce strangulation. Unconsciousness,

---

* VON PETTENKOFER UND ZIEMSSEN: Handbuch der Hygiene, etc. II Th. 4 Abth.,
p. 30.
† See Chap. I, p. 16.

convulsions and asphyxia rapidly succeed. Paralysis of the sphincters and of groups of other muscles are often present. The pulse is at first somewhat increased but soon becomes slower. The respiration is slow and stertorous, and the temperature falls from 3°-4° F. Glycosuria often occurs. If death does not occur in the attack, the patient frequently suffers from great depression, both physical and mental, loss of appetite, constipation, and various paretic conditions.

The slow or chronic form of poisoning by carbonic oxide is characterised by headache, dizziness, slow pulse and respiration, nausea and sometimes vomiting and purging. Loss of memory, and diminution of mental activity are also said to be effects of the continued inhalation of air charged with carbonic oxide.

Carbonic acid gas is found as one of the constituents of the 'choke-damp' in mines. There is reason to believe that this is often the source of ill-health and death in miners, even where the symptoms of acute carbonic acid poisoning are not present. Hon. Andrew Roy[*] says that it 'is more insiduous than direct in its operations, gradually undermining the constitution and killing the men by inches.' Difficulty of respiration and weakness are the only symptoms calling attention to the pernicious effects of the gas. Where, however, the proportion of carbonic acid is large, acute poisoning occurs. This is manifested by the following symptoms: loss of consciousness and of the power of voluntary motion. In some cases there are convulsions; in others the above symptoms are preceded by difficult respiration, headache, depression, drowsiness, or psychical excitement. Recovery usually soon follows after removing the patient into a purer atmosphere.

Vintners, distillers, brewers and yeast makers are said to suffer from the effects of carbonic acid occasion-

---

[*] Third Annual Report State Mine Inspector of Ohio. Quoted in Beck's Hygiene and Public Health, Vol. II., p. 243.

ally, but serious results from this cause are probably very infrequent.

Sulphuretted hydrogen when present in the air in large proportion, as for example, in privy vaults, cesspools, and sewers may produce serious or fatal poisoning. Formerly, when vaults were cleaned in the primitive way, these accidents were frequent, but at the present day, owing to improved methods of removing excreta, they are comparatively rare. The precautions advised in a preceding chapter* should be borne in mind when it is necessary for workmen to enter such places.

The gases resulting from the *putrid decomposition of organic substances*, such as are found in tanneries, glue and soap works, and similar industries, are popularly believed to give rise to various diseases. There are no observations on record, however, to show that such is the case. As a matter of fact, the workmen engaged in the industries mentioned, seem to be exceptionally healthy, and to resist to a considerable degree the ravages of phthisis and of epidemic diseases.

Bisulphide of carbon is used in the arts principally in the process of vulcanizing India rubber, and for extracting oils from seeds and fatty bodies. The constant inhalation of the vapor of bisulphide of carbon produces a train of symptoms to which attention was first attracted by Delpech, in 1856. The symptoms have been observed frequently since that time. The following account is from Hirt :†

'Some days, or even weeks or months after beginning this occupation, the workmen complain of a dull headache, becoming more severe toward evening. This symptom is soon followed by joint-pains, formication, and itching on various parts of the body. A more or less troublesome cough is present, but is not accom-

* Chapter I., p. 21.
† Op. cit., p. 66.

panied by any characteristic sputa. The respiration is regular, the pulse somewhat increased in frequency. During this time certain individuals exhibit a marked exaltation of their intellectual powers, they talk more than formerly, and show an interest in matters in which they at other times show no concern. There is, however, very rarely distinct mental disease. The sexual desires are increased in both sexes, menstruation becomes irregular, and the urine possesses a faint odor of bisulphide of carbon. In this manner several weeks or months pass away. Very gradually the physical exaltation disappears, and a profound depression, melancholy and discouragement succeeds, coupled with which is often loss of memory. Vision and hearing become less acute, and the sexual activity is completely destroyed. Anesthetic spots appear on various parts of the body, and numbness of the fingers prevent the workman from performing any fine work.'

The disease never proves fatal, but the normal condition of the individual is rarely re-established when the disorder has advanced to the extreme stages mentioned.

Iodine and bromine vapors when inhaled by workmen engaged in their preparation produce symptoms of poisoning which are sometimes very serious. Acute iodic intoxication consists in severe laryngeal irritation, headache, conjunctivitis, and nasal catarrh. Occasionally there is temporary loss of consciousness. Chronic iodic cachexia is often found among the workmen. In certain cases atrophy of the testicles and gradual disappearance of sexual power has been observed. In the manufacture of bromine, a form of bronchial asthma has been observed among those engaged in the establishment. No symptoms corresponding to those of chronic iodism have been observed among the workmen in bromine.

The inhalation of the vapors of *turpentine* produce in a considerable number of those constantly exposed to them, diseases of the respiratory organs, beginning with cough, and, at times, resulting in consumption. In other cases derangements of the digestive organs, strangury and in a few cases bloody urine, have been observed. Nervous disturbances are rare after the inhalation of turpentine, and are limited to headache, roaring in the ears, or flashes of light before the eyes.

Petroleum vapor when inhaled in a concentrated state produces symptoms similar to those of anesthetics. When exposed for a long time to diluted petroleum vapor, workmen sometimes suffer from chronic pulmonary catarrhs, or from nervous derangements. Among the latter are disturbances of mental activity, loss of memory, giddiness and headache. These symptoms are however rare. More frequent are pustular or furuncular affections of the skin, which are due probably to the direct irritant effect of the vapor.

Lead-poisoning is one of the most characteristic diseases of artisans. It attacks workmen engaged in roasting and smelting of lead ores ; in the manufacture of white and red lead and of lead acetate and chromate; in type making, in painting, and in short in all occupations in which the workman is compelled to inhale the vapor or dust of lead, or in which it is conveyed in some manner to the digestive organs. It is believed also that it can be absorbed by the skin and produce its poisonous effects upon the economy. The average duration of life in the roasting and smelting furnaces is 41 years; of painters, as shown by table I., 45.07 years. Of the latter 75 per cent. are attacked by one of the forms of lead-poisoning, colic being most frequent. In the manufacture of white lead, more than half of the workmen suffer from lead-poisoning during the first year, lead colic being present in 60 per cent. of all the cases.

In most sugar of lead manufactories 60 per cent. of all the operatives constantly suffer from some form of lead-poisoning.

Poisoning has also been observed in workmen engaged in the manufacture of various pigments of which the acetate of lead is the base (e. g. lead chromates). Among type founders, the symptoms of lead-poisoning are not very rare, and even compositors sometimes suffer from lead-poisoning. In the latter case the lead must be absorbed through the skin in order to produce its effects.

The various forms in which lead-poisoning affects the individual are the lead cachexia, manifested by loss of weight, discoloration of the skin, the characteristic blue line along the gums, diminution of the salivary secretion, a sweetish taste and offensive odor of the breath; then lead colic, the features of which are well-known; lead paralysis, the characteristic 'wrist drop,' which requires prompt and intelligent treatment, otherwise permanent atrophy of the affected muscles often takes place. Among other nervous manifestations of the poison is a painful affection of the lower extremities, attacking joints and flexor muscles, and remittent in character. At times anesthesia of the skin of the head and neck is present. In rare cases, serious mental derangement occurs. Other grave nervous lesions, such as the so-called saturnine hemiplegia and tabes are happily extremely rare among the workmen in the metal at the present day.

Mercurial poisoning is frequent among the artisans who work in the metal. The smelters of the ore suffer severely and in a large proportion of the entire number employed. Their average age at death is forty-five years. Mirror makers suffer most severely of all the artisans who come in contact with the vapors of the metal. It is beyond question that the confinement in badly ventilated work-rooms is largely responsible for the poisonous effects of the metal upon this class. The special forms in which the poisonous effects are manifested in mirror makers are salivation, mercurial tremor, and nervous erethism, but in addition a very large proportion suffer from pulmonary consumption. It is stated that 71 per cent. of the total deaths among mirror makers (those who coat the glass with the mercurial alloy) are from phthisis.

Among women the symptoms are aggravated, and abortion frequently occurs. Of the children of women suffering from mercurial poisoning born living at term, 65 per cent. die within the first year.

Fire gilders, fulminate makers, and physical instrument makers not infrequently suffer from the deleterious effects of inhaling the vapor of mercury. Hatters are also liable to a considerable extent to the poisonous effects of the metal.*

It has been found that upon sprinkling the floor of the work-room of mirror makers with aqua ammonia, so as to impregnate the atmosphere with ammonia, the bad effects of mercury on the system were markedly diminished. Care must be taken, however, not to use the ammonia to excess, otherwise the diseases caused by this agent may attack the workmen.

Zinc or copper vapors, or possibly a combination of the two, given off from brass, which is an alloy of these metals, produces a peculiar train of symptoms known as 'brass-founder's ague.' The symptoms are described by Hirt, who has suffered from two attacks of the affection himself, as follows :† 'A few hours after attending the process of brass casting, one notices a peculiar, uncomfortable sensation over the whole

---

* L. DENNIS: Hatting, as Affecting the Health of Operatives. Report New Jersey State Board of Health, 1879.—Connecticut State Board of Health, 1883.

† Op. cit., p. 122.

body. More or less severe pains in the back, and general lassitude, cause a discontinuance of the ordinary occupation. While the pains appear now here, now there, and are extremely annoying, no changes in the pulse or respiration are noticeable. In a short time, however, usually after the patient has taken to the bed, chilliness comes on which soon increases to a decided rigor, lasting fifteen minutes or longer. In the course of an hour or less, the pulse now reaches a rapidity of 100–120 beats per minute. A tormenting cough, combined with a feeling of soreness in the chest comes on. In consequence of the repeated acts of coughing, the increasing frontal headache produces exceeding discomfort. Soon, however, usually after a few hours, the height of the attack is reached, free perspiration indicates the stage of defervescence and during the gradual diminution of the symptoms, the patient falls into a deep sleep lasting several hours. On awaking, a slight headache and lassitude only remain as reminders of the attack.

It is said that about 75 per cent. of the workmen in brass foundries are attacked by this affection; the attack is liable to be repeated at every exposure.

A chronic form of poisoning is said to occur among zinc smelters, after following their occupation for ten to twelve years. It consists of hyperesthesia, formication and burning of the skin of the lower extremities, soon followed by alteration in the temperature and tactile sensation, and diminution of the muscular sense. Paresis of the lower extremities sometimes comes on. The disease has not yet been sufficiently investigated.

Anilin vapor is exceedingly poisonous when inhaled in a concentrated state. Hirt describes an acute form which usually results fatally. 'The workman falls suddenly to the ground, the skin is cold, pale, the face is cyanotic, the breath has the odor of anilin, the respiration is slowed, and the pulse increased. The sensation, diminished from the beginning of the attack, gradually entirely disappears, and death follows in a state of deep coma.'* There is a milder form which comes on after several days of exposure. It is characterised by laryngeal irritation, diminution of appetite, headache, giddiness, great weakness and depression. The pulse is rapid, small and irregular. Respiration is little altered. There is decrease of sensibility of the skin. Convulsions may occur, but are usually of short duration.

The chronic form of anilin poisoning is characterised by three sets of symptoms: those affecting the central nervous system, the digestive tract, and the skin. Among the first are lassitude, headache, roaring in the ears, disturbances of sensation and motion of greater or less degree.

The digestive derangements consist in eructations, nausea and vomiting.

The cutaneous lesions are eczematous or pustular eruptions, and sometimes round, sharply circumscribed ulcers with callous borders.

There is no trustworthy evidence that in the manufacture of *anilin colors* poisonous symptoms are produced in the workmen.

II.—DISEASES DUE TO THE INHALATION OF IRRITATING OR POISONOUS DUST.

The inhalation of air containing particles of organic or inorganic matter has long been accepted as a cause of certain special diseases of artisans. The diseases so caused are usually limited to the pulmonary organs, and consist of acute and chronic catarrh,

* Op. cit., p. 127.

emphysema of the lungs, pneumonia, interstitial inflammation of the lungs, the so-called fibroid phthisis
or pulmonary cirrhosis.

Coal dust is inhaled by coal miners, charcoal burners, coal handlers, firemen, chimney sweeps, foundrymen, lead-pencil makers, etc.   Chronic bronchial
catarrhs are most frequent, while phthisis and emphysema are almost absent from the list of diseases affecting these workmen.   The table on page 166 shows that
the expectation of life of foundrymen, furnace-men,
firemen and chimney sweeps is much below the average.

Metallic dust is inhaled by blacksmiths, nailers,
cutlers, locksmiths, file-cutters, cutlery and needle polishers, etc.   While in this class of workmen cases of bronchitis and pneumonia are relatively frequent, much the
largest proportion suffer from phthisis.   A table com-
piled by Hirt shows that out of the total number of
sick in the different classes of workmen the cases of
phthisis were :

> 62.2 per cent. for file cutters,
> 69.6 "   "   " needle polishers,
> 40.4 "   "   " grinders,
> 12.2 "   "   " nailers.

The Massachusetts table gives the average duration
of life for blacksmiths at 53.26 years, of nail makers
at 41.49 years, and of cutlers at 39.21 years.   The
needle polishers at Sheffield, as stated above, (page 168)
have only an average duration of life of 30.66 years.
In this work and that of grinding knives, scissors and
similar articles, the metallic dust is mixed with mineral
dust, (particles of silica from the grindstone).   This
mixture seems to be much more deleterious than metallic dust alone, as shown by the shorter average duration
of life and the enormous percentage of cases of consumption.

Mineral dust is inhaled by the workmen in a large number of different industries. The grinders in the ground-glass factories suffer most severely. Hirt found the average duration of life in grinders who began this occupation after their twenty-fifth year to be 42.50 years, while in those who began at the age of fifteen, the average duration was thirty years.

Mill stone cutting is also a very dangerous occupation. Peacock* gives the average age at death of these workmen at 24.1 years. Stone-cutters generally, suffer frequently from phthisis, probably largely in consequence of the constant inhalation of the mineral dust produced during their work. The Massachusetts table gives the average age at death of these workmen at 40.90 years, while Hirt's table gives a much lower age, namely 36.3 years. Potters and porcelain makers are exposed to similar dangers from their occupation, but to a much less degree. The table on page 166 gives the average age at death at 56.67 years, rather a high average.

Slaters and workmen in slate quarries suffer in a large proportion of cases from chronic pneumonia, and die at a comparatively early age.

Masons and carpenters have an average duration of life of 50.33 and 53.33 years respectively. One-third of all the diseases from which they suffer, affect the respiratory organs.

Gussenbauer has reported a series of cases of a peculiar inflammatory affection of the diaphyses of the long bones, in the artisans who are engaged in the manufacture of pearl buttons.

Gem finishers are exposed not only to the inhalation of dust, but to poisonous gases (carbonic oxide) and vapors, (lead). The proportion of sickness among them is very high.

---

* Quoted by MERKEL, in VON PETTENKOFER UND ZIEMSSEN. Handbuch der Hygiene, IITh. 4 Abth. p. 197.

VEGETABLE DUST.

The workmen compelled to inhale vegetable dust are those who work in tobacco, cotton operatives, flax dressers, paper makers, and weavers, wood turners, millers and laborers in grain elevators.

Workmen in tobacco usually suffer within a few weeks after beginning work, from a nasal, conjunctival and bronchial catarrh which soon passes off, as the mucous membranes seem to become accustomed to the irritation. Nausea is also frequent at first, due prob- ably to the absorption of small quantities of nicotin. Females exposed to tobacco dust usually suffer from digestive and nervous troubles. They are said also to abort frequently. Dr. R. S. Tracy,* as a result of his observations among the cigar makers in New York, states that the fecundity of these people is much less than the average. 325 families visited had only 465 children, an average of 1.43 to each family. Dr. Tracy is inclined to attribute this to the frequent abortions that occur among the females exposed to the inhalation of tobacco dust. According to the Massachusetts table, cigar making is an unfavorable occupation, the average age at death being 38.36 years.

Cotton operatives, flax dressers, weavers, and work- men in paper mills are subject to various diseases of the respiratory organs. Coetsem, as long ago as 1836, described a peculiar pulmonary affection among cotton operatives, which he termed *pneumonie cotonneuse.* The observation does not seem to have been verified by others ; at all events, the author is unable to find any other record of a similar affection in the literature of the subject. Among weavers the mortality from phthi- sis is comparatively high. Among paper makers Hirt found an average duration of life of 37.6 years. The

---

* Buck's Hygiene and Public Health. Vol. II, p. 42.

people who sort rags are liable to a fatal infectious disease, 'rag-sorter's disease,' (Hadernkrankheit*) which resembles in all respects, and is probably nothing less than anthrax. No cases have been reported in this country, but as the importation of rags from abroad is carried on to a considerable extent, no apology is believed to be necessary for calling attention to it. The 'wool-sorter's disease,' to which attention has recently been called in England, is doubtless similar in its nature.

Millers suffer in a large proportion of cases from pulmonary affections, especially bronchial catarrh and pneumonia. According to Hirt, 20.3 per cent. of all the diseases of these workmen are pneumonias, 9.3 per cent. bronchial catarrhs, 10.9 per cent. phthisis, and 1.9 per cent. emphysema. The average duration of life, is 45.1 years. The Massachusetts table gives 57.14 years, a very much more favorable exhibit.

The laborers in grain elevators are compelled to inhale a very irritating dust which causes acute and chronic catarrhs of the respiratory organs. Dr. T. B. Evans, of Baltimore, has reported a series of cases of catarrhal pneumonia in these workmen, which were characterised by some peculiar features. Brush making, according to the statistics of Hirt, is a very dangerous occupation. Nearly one-half of the deaths among brush-makers are from phthisis, due, in great measure to the inhalation of the sharp fragments of bristles produced in trimming the brushes. In the Massachusetts table the average duration of life is given at 43.11 years.

III.—DISEASES DUE TO THE ABSORPTION OR LOCAL ACTION OF IRRITATING OR POISONOUS SUBSTANCES.

Arsenic is used in the manufacture of green pigments, and for various other purposes in the arts. In

---

* See Article by SOYKA: Realencyclopædie d. ges. Heilk. Bd VI, p. 156.

the preservation of furs, and in taxidermy it finds extensive use.   In the preparation of the pigment known as Paris green, the workmen are frequently entirely covered by a layer of the poisonous salt.   The poisonous symptoms occur in consequence of the absorption of the poison through the skin, or from its local action, and but rarely on account of inhalation of vapors or dust in which it is contained.   The most marked symptoms are chronic gastric catarrh, superficial erosions in the mouth, dry tongue, thirst and a burning sensation in the throat.   These symptoms may continue for months or even years and gradually produce a complete breaking down of nutrition and the vital powers. Violent itching skin eruptions, of an eczematous character are not infrequent complications of the internal symptoms.

Phosphorus produces two classes of effects in persons subjected to its influence.   The milder effects are produced by the inhalation of the fumes of the substance, and are limited to digestive disturbances, and diseases of the pulmonary organs.   The severer symptoms are only observed among the employés in match factories, and are due to the local action of the phosphorus upon the tissues affected.

The characteristic disease produced by phosphorus is a painful periostitis of the lower or upper jaw.   The limitation of the affection to this locality is believed to be due to the action of the phosphorus dissolved in the saliva.   The fact that the lower jaw with which the saliva comes more thoroughly in contact is most frequently affected, seems to indicate that this view is the correct one.   The disease begins, on an average, five years after the beginning of the employment.   Hirt estimates the proportion of employés in match factories attacked at 11 to 12 per cent.   The first symptom of the disease is tooth-ache, soon extending to the jaw.

The cervical glands swell up, the gums become reddened and spongy, abscesses form about the diseased teeth from which large quantities of thin, offensive pus are discharged. Examination with a sound reveals carious, nodulated bone. The cheeks become swollen, erysipelatous, and may suppurate and discharge pus externally.

The destruction of the soft tissues continues until resection of the jaw is finally undertaken and the disease checked by surgical interference, and removal of the patient from the influence of the pernicious substance.

In the manufacture of *quinine*, a troublesome eczema is caused in about ninety per cent. of the employés. It seems to be due to emanations given off from the boiling solutions. It begins with intense itching, followed by swelling and the formation of vesicles, which soon burst and form crusts. There is considerable fever when the swelling is great. It is said that blondes are more frequently affected than those of dark complexion. The disease soon disappears if the work is given up.

The workmen engaged in the manufacture of *bichromate of potassium* are said to suffer from an ulceration of the nasal mucous membrane very similar to that already described as due to the vapors of hydrochloric acid (p. 170). Rapidly spreading, deep ulcers are also said to form, if the bichromate comes in contact with abraded surfaces of the skin.

The *strong alkali* handled by tanners frequently produces fissured eczemas of the hands which are painful and often difficult to cure.

The workmen in *petroleum refineries* frequently suffer from acneiform or furuncular eruptions.

Among glass blowers, syphilis is frequently communicated by an infected mouth-piece which is used by the men in turn.

IV.—DISEASES DUE TO EXPOSURE TO ELEVATED OR VARI-
ABLE TEMPERATURE OR ATMOSPHERIC PRESSURE.

Cooks and bakers are exposed almost constantly to
a high temperature, which produces an unfavorable
influence upon health, and predisposes them to diseases
of various kinds. The Massachusetts table shows that
cooks have a much shorter duration of life than bakers,
although the statistics of both trades are unfavorable.

The prevailing diseases among cooks and bakers
are rheumatism and eczematous eruptions, generally
confined to the the hands, forearms and face.

Blacksmiths, founders, and firemen suffer from the
intense heat to which they are exposed, in addition to
the inhalation of coal dust, as has already been pointed
out (p. 180). The stokers in the engine-rooms of
steamships suffer especially from the excessively high
temperature to which they are subjected by their
occupation. A form of heart-weakness, described by
Levick as 'fireman's heart,' is prevalent among them.

Sailors, farmers, coachmen, car drivers and team-
sters are subjected to stress of weather, changes of
temperature and storms. They suffer frequently from
rheumatism, acute bronchitis, pneumonia and Bright's
disease. Car drivers are said also to suffer from painful swelling of the feet, varicose veins and ulcers, and
mild spinal troubles.*

Sunstroke is not confined to any class of artisans,
but persons who perform hard labor, especially in a
confined atmosphere, suffer most frequently.

The effects of *compressed air* on workmen in tunnels
and deep mines has already been referred to.† The
most serious symptoms occur, not when the individual
is subjected to the increased pressure, but when the
pressure is too rapidly diminished.

---

* A. McL. Hamilton in Report New York Board of Health, 1873, p. 444.
† Chapter I, p. 8.

V.—DISEASES DUE TO THE EXCESSIVE USE OF CERTAIN ORGANS.

The prevalent belief that the over use of the intel-lectual faculties is a frequent cause of mental disease is not borne out by facts. Men and women who per-form an amount of mental work regarded by most persons as excessive, have in spite of this, a long duration of life. There are no exact statistics upon this subject, but Casper a half century ago made the following estimate of the average duration of life among professional men : clergymen live 65 ; merchants, 62.4 ; officials, 61.7 ; lawyers, 58.9 ; teachers, 56.9, and physicians, 56.8 years. In the table given on page 166, the figures are somewhat less favorable, although cor-responding in general with those of Casper. Hence, it is seen that of professional men, those whose occupa-tion compels the exercise of high mental powers, have a higher duration of life than any other class, except farmers and mechanics engaged actively out of doors. Those professional occupations only which necessitate a more or less irregular mode of life and frequent sub-jection to physical exhaustion and dangers from contagious disease, such as physicians and journalists, make an unfavorable showing in the statistics. The proposition may be laid down that it is not mental *activity*, however great, but mental *worry* that tends to the abbreviation of life.

The occupation of a tea-taster is said to produce a peculiar nervous condition, manifested in muscular tremblings, etc., which compels the individual to give up the work in a few years.

Persons who test the quality of *tobacco*, an occu-pation corresponding to that of tea-taster, are said to suffer often from nervous symptoms, which may include amaurosis and other grave affections.

Those persons who are compelled to use their eyes constantly upon minute objects frequently suffer from defective vision. So engravers, watchmakers, seamstresses, are liable to near sightedness, amaurosis, and irritation of the conjunctiva. Public speakers and singers frequently suffer from catarrhal, or even paretic conditions of the throat, which usually disappear on relinquishing the occupation for a time.

Telegraph operators and copyists suffer from a peculiar convulsive affection of the fingers, called 'writer's cramp.' Performers on wind instruments are liable to pulmonary emphysema, on account of the pressure to which the lungs are frequently subjected. Boiler-makers often suffer from deafness, in consequence of their constant existence in an atmosphere in a state of continual violent vibration. The affection is generally recognized as 'boiler-maker's deafness.'

VI.—DISEASES DUE TO A CONSTRAINED ATTITUDE AND SEDENTARY LIFE.

It is probable that the large mortality and morbility rate of persons whose occupations keep them confined within doors are due, next to the defective ventilation, to the constrained attitude which most of them necessarily assume. Thus, carvers, book-binders, engravers, jewellers, printers, shoe-makers, book-keepers and cigar-makers all have a low average duration of life. It is found likewise that many of these artisans suffer most from pulmonary and digestive troubles; among the former being phthisis, and among the latter constipation, dyspepsia and hemorrhoids.

VII.—DISEASES FROM EXPOSURE TO MECHANICAL VIOLENCE.

It will be seen by reference to the table on page 166, that all persons whose occupations involve an intimate contact with machinery, and in the pursuit of which accidents frequently happen, have a short duration of life. Persons liable to these dangers are machinists, operatives in factories, workmen in powder mills, baggage masters, brakemen, drivers, engineers, firemen, and other workmen on railroads. Aside from the diseases to which some of these classes are liable, in consequence of exposure to variable atmospheric conditions, the grave accidents to which they are so frequently exposed, render their occupations extremely dangerous. Brakemen on freight railroads, for example, are classed by insurance companies as the most hazardous 'risks,' and some companies refuse to take them at all. The table on page 166 tends to confirm the conclusion of the insurance companies for excluding the class of 'students' which, for manifest reasons, cannot be used as a comparison, brakemen have the shortest average duration of life of all the occupations noted in the table.

[The student is referred for more complete information on the subjects considered in the foregoing chapter to the following works:

L. HIRT: Die Krankheiten der Arbeiter. EULENBURG : Handbuch der Gewerbehygiene. LAYET: Hygiene des Professions et des Industries].

CHAPTER X.

MILITARY AND CAMP HYGIENE.

THE subjects embraced in this chapter can be most conveniently arranged under the following heads :—

I. The Soldier ; his Training, Food, Clothing and Shelter.

II. The Diseases to which Soldiers are especially liable.

III. Civilian Camps.

I.—THE SOLDIER AND HIS TRAINING.

The relations existing among different nations at the present time, require that a standing army of greater or less number be maintained by each for the common safety. This being the case, it needs no argument to prove that such an army should be composed of the best material available in order that it may be depended upon for defense or offense when necessity demands that it should be called into active service.

Hammond says with truth[*] that 'a weak, malformed, or sickly soldier is not only useless, but a positive incumbrance' to an army. It is of the first importance, therefore, to exclude from the military service by a vigorous physical examination, all individuals whose physical condition is defective, who are either suffering from, or predisposed to disease.

The foremost authorities on military hygiene are agreed that no recruit should be enlisted for actual service before the twentieth year. In the English army the lowest age at present is nineteen years, in Germany twenty years, and in the United States twenty-one years.

---

[*] Hygiene, Philadelphia, 1863. p. 19.

The limit of age upward is forty-five years, except in cases of re-enlistments. The height of recruits must be at least five feet four inches ; minimum chest measurement 30 inches, with two inches expansion, and weight from 120 to 180 pounds. In the cavalry service the maximum weight is 165 pounds. Every recruit must be vaccinated before enlistment.

The physical examination of recruits before enlistment must be made by a medical officer, whose decision, in the United States army, is definitive. In the German army the decision of the medical officer is not final, but subject to revision by the recruiting officer, who may, if he sees fit, overrule the medical officer's opinion and enlist a man who has been decided to be unqualified for the military service. In this, and various other respects, such as pay, rank, and effective power the Medical Staff of the United States army has many advantages over that of most foreign armies.

II.—THE FOOD OF THE SOLDIER.

Thé army ration of the United States, which is given below, approaches moderately near to the standard quantity of food for a healthy male adult.* The daily allowance for each soldier is as follows :

```
12  oz. pork or bacon, or
20   "  fresh beef.
16   "  hard bread.
2.4  "  beans or peas, or
1.6  "  rice or hominy.
1.6  "  green coffee, or
1.28 "  roasted coffee, or
 .24 "  tea.
2.4  "  sugar.
 .6  "  salt.
 .04 "  pepper.
 .04 qts. vinegar.
```

Although the food allowance in the United States army is greater than in the British army, the medical

---

* See Chapter III, p. 59.

officers of our army insist that the ration furnished is insufficient in quantity as well as not sufficiently varied to fulfil the demands upon it during active service.

The money value of each of the above articles in the ration is fixed by the government and may be drawn instead of certain of the articles, and other articles of food purchased, and thus the dietary varied. The money so drawn constitutes what is known as the 'company fund.' In the hands of a judicious commanding officer, the company fund can be made a source of great benefit and comfort to the men, but that it is at times mismanaged or misapplied is well-known to army surgeons.

Aside from the insufficient quantity of food furnished to soldiers, the cooking, especially in temporary camps, is often defective and causes digestive derangements and consequent innutrition. A good cook should be attached to every company.

THE CLOTHING OF THE SOLDIER.

The clothing of the United States soldier is tolerably well adapted to its uses. It is generally well-made of good, serviceable material. The only exception that can be made with reason is that the foot-gear is not made to individual measure, and hence peculiarities of shape of the feet, cannot be taken account of. For this reason painful affections of the feet are of frequent occurrence, due to ill-fitting boots or shoes.

When on a march, the soldier carries his extra clothing packed in a knapsack, and strapped upon the back. His blankets and great coats are rolled into a cylinder and strapped upon the knapsack. The weight each soldier has thus to carry, in addition to his arms and ammunition amounts to considerable. There is reason to believe that the pressure produced by the straps of a heavy knapsack, may cause not only discomfort, but actual disease. It is believed by many officers that the knapsack could be abolished with advantage, and the extra clothing rolled up in the blanket, or a water-proof sheet, and slung over the left shoulder.

THE DWELLING OF THE SOLDIER.

Soldiers are generally housed either in barracks, huts, or tents. The former are usually the habitation of the soldier in garrisons or permanent camps, while huts and tents are used for the purpose of sheltering the occupants of temporary camps.

Barracks.—A military barrack is a one-story building constructed of stone, wood, or iron, or a combination of these materials. The general plan of the barrack comprises a large room for the beds of the soldiers, one or more smaller rooms for the non-commissioned officers of the company or squad, and a wash-room. The sleeping room of the soldier is also his living or day room. It is evident therefore that sufficient air space and good ventilation must be provided, if the soldier's health is to be maintained. In England, 600 cubic feet is recommended for the initial air space. In the new barracks constructed in France according to the plans of M. Tollet, 770 cubic feet are allowed to each occupant.

The special points of distinction of the system of Tollet, of which Schuster says that 'to it belongs the future of barrack construction' are: The frame of the building is of light iron ribs; the interspaces are walled up with bricks or stone; the roof is slate; the ceiling is arched and all corners are rounded to prevent lodgment of dust.

Ventilation is provided by openings in the walls at the edge of the roof for the entrance of fresh air, and ridge ventilators. In France, barracks have been built according to Tollet's system at Bourges, Cosne,

Macon and Autun. Although occupied but a short time, it appears that the health of the soldiers remains much better in them, than in the barracks constructed on the old style. The system would seem also especially to lend itself to the construction of hospitals. The wash and bath-rooms of the barracks should be so arranged as to encourage the soldier to cleanliness. Where the only lavatory in a barrack is, as the author has seen it, an open porch, men are not likely to spend much time in cold weather in washing their faces and hands, to say nothing of the rest of their bodies.

The kitchen and dining room should be detached from the building serving as quarters; otherwise the odors of the cooking will pervade the building.

The sinks or latrines should be placed at some distance from the quarters and kitchen, and out of the line of prevailing winds. The writer has personal knowledge of a permanent military post within a few miles of the city of Washington, where, only a few years ago, (and for aught known to the contrary, at the present day) 'the rear', or place of depositing excrement, was about 75 yards distant from the kitchen and men's quarters, and directly in line, on the windward side, with the prevailing winds!

Before erecting any buildings it is of course necessary to endeavor to secure a clean and dry subsoil. Attention is called to the principles underlying the construction of dwellings, Chap. VI.

Tents and Huts.—The tents used in the army are the hospital tent, the officers' wall tent, the A tent and the shelter tent, which is a modification of the last. The conical or Sibley tent, which was frequently seen in camps in the early part of the war between the States, has gone out of use. Soldiers give the preference to the shelter tent, which is light, each man's piece weighing only two pounds six ounces. Two pieces being joined together by buttons and button holes, and thrown over a ridge pole supported upon uprights, and the four corners fastened to pegs driven into the ground, form a tent 4 feet high, 5 feet 6 inches long, and having a spread at the base of between 6 and 7 feet. Such a tent will form a comfortable shelter for two men, unless there should be strong winds or driving rains, when the ends could be closed by blankets, brush, or an extra piece of shelter tent. The uprights and ridge are steadied by short guy ropes, one of which is furnished with each piece of the tent.

In winter, especially when camps of more or less permanence are formed, the soldiers usually build log huts. The interstices between the logs are plastered up with mud or clay, and the roof is formed of canvas, generally several pieces of shelter tent joined together.

The ground around the tent or hut should be trenched in order to carry off the rainfall.

Cleanliness within and around tents or huts is of the first importance, and should be enforced in all camps by the proper authority. Military authorities have long since learned that in the matter of cleanliness of body, clothing or surroundings no dependence can be placed upon the soldier. Frequent and thorough inspections will alone secure proper cleanliness.

CAMP DISEASES.

The soldier's profession has been aptly characterised by Ruskin as 'the trade of being slain.' In the late war between the States, the total deaths of the Federal army numbered 279,659; over ten per cent. of the entire number of enlistments. Of this number however, 186,216, or nearly two-thirds died from disease, while the remaining 93,443, a small fraction over one-third, were killed in battle or died from the effects of wounds.

Diarrhœa and Dysentery.—The most fatal diseases of camp life, especially in time of war, are diarrhœa and dysentery. The statistics of the Federal army during the late war are given in the following table.*

Table I.—*Total Deaths from Diarrhœa and Dysentery in the U. S. Army, from May 1st, 1861, to June 30th, 1866.*

	WHITE TROOPS, FROM MAY 1ST, 1861, TO JUNE 30TH, 1866.		COLORED TROOPS, FROM JULY 1ST, 1863, TO JUNE 30TH, 1866.		TOTAL.	
	CASES.	DEATHS.	CASES.	DEATHS.	CASES.	DEATHS.
Acute Diarrhœa,	1,155,226	2,923	113,801	1,368	1,269,027	4,291
Chronic Diarrhœa,	170,488	27,558	12,098	3,278	182,586	30,836
Acute Dysentery,	233,812	4,084	25,259	1,492	259,071	5,576
Chronic Dysentery,	25,670	3,229	2,781	626	28,451	3,855
Total, - -	1,585,196	37,794	153,939	6,764	1,739,135	44,558

Owing to the fact that a considerable number of deaths were reported without assigning any cause, Dr. Woodward estimates the total number of deaths from the above diseases at 57,265, or, in the proportion of one death from diarrhœa and dysentery to three and one-half deaths from all diseases. Among the prisoners of war held by the Confederate States in Andersonville prison, where tolerably complete records were kept, more than half the total deaths were from diarrhœa and dysentery, while the ratio of deaths to cases of the above two diseases was a fraction over seventy-six per cent. This frightful mortality from these two diseases, both in the prisons and among the armies in the field is principally due to the insanitary conditions surrounding the soldiers. Where the demands of hygiene were especially ignored; where the food was bad in quality, or badly cooked; the water impure; the soil polluted by excreta and other filth; where the men were exposed to stress of weather or to a paludal atmosphere—under these conditions, the above diseases of the intestines prevailed in their greatest extent and most fatal degree.

---

* Medical and Surgical History of the War. Second Medical Volume, p. 2.

Malarial Fevers.—The diseases due to the paludal poison are exceedingly frequent among soldiers encamped in malarial sections. During the civil war a very pernicious form of malarial fever received the designation of the locality in which it prevailed, and passed into the literature under the name of 'Chicka-hominy fever.' While malarial diseases were largely represented in the morbility reports during the war, the most serious results of the influence of the malarial poison were manifested in its effects upon patients sick with other diseases. Thus, typhoid fever, dysentery or pneumonia in a patient saturated with malaria was very much more serious than where this complication was absent. In the malarial regions in the interior of the country, the Mississippi Valley, and the Southern portion of the Western territories, malarial fevers are among the most prevalent camp diseases. Greater attention in locating camps, and care devoted to drain-ing the subsoil and maintaining a low level of the ground water, would doubtless result in improvement in the sickness-rate from this cause in the army.

Typhoid Fever.—Typhoid fever is prevalent in camps and garrisons. As it may be propagated through the medium of infected discharges of typhoid patients, it will readily be perceived that neglect of the precaution of promptly disinfecting such discharges will almost inevitably result in spreading the disease, either by direct inhalation of effluvia from the patient or excreta, of pollution of the drinking water, or by contamination of the soil and subsequently of the atmosphere, by the intestinal discharges of the patient.

Phthisis.—Especially among troops in barracks, phthisis is a very fatal disease. Formerly the mortality from it was very heavy. Recent improvements in the hygiene of military posts, and greater care in selecting recruits have very greatly diminished the death-rate from phthisis among soldiers. Acute pulmonary affections, such as bronchitis, pleurisy and pneumonia are comparatively frequent in camps, being due to exposure.

Typhus Fever and Scurvy.—These two diseases are at the present day comparatively rare as camp diseases. They break out, however, on every occasion when the laws of hygiene are violated by permitting overcrowding, overwork, and under-feeding. This is almost certain to occur during war, and hence, either fully developed scurvy, or a scorbutic taint are almost constant accompaniments of an army in the field. Among the allied armies in the Crimea, and in the Federal army during the war, scurvy and typhus fever claimed a considerable share in the mortality.

Purulent Conjunctivitis.—This affection of the eyes is frequent among soldiers. It has even been supposed to be peculiar to soldier life, and has hence been termed 'military ophthalmia.' It is contagious, and is probably most often spread by the use of basins and towels in common. It is not merely annoying, but is a very grave affection, often causing perforation of the cornea and destruction of vision. The military surgeon should be on the lookout for it, and promptly isolate those infected.

CIVILIAN CAMPS.

The camps of civil life, whether established for the purpose of furnishing a refuge to the inhabitants of cities invaded by epidemic disease, yellow fever or cholera, or whether for religious purposes (camp meetings), or for recreation (hunting and fishing camps, etc.), should be organised and managed on the same principles as the military camp. The site should be selected with judgment—a clean, dry soil, abundance of wood and water being requisites for a healthy camp.

A superintendent, or officer of the day should be appointed, whose duty it is to carefully inspect the camp daily, and compel the prompt removal of all filth and offal from the immediate vicinity. Cleanliness of person, clothing and household is as important while 'roughing it' in camp, as at home. Singularly this is very often forgotten by very intelligent people.

The advantage of a well-administered refugee camp in case of yellow fever epidemics has been clearly shown by the brilliant success of the depopulation of Memphis during the epidemic of 1879. This experiment deserves imitation.

[The following works on Military and Camp Hygiene should be studied in connexion with this chapter:

SMART: The Hygiene of Camps; in BUCK's Hygiene and Public Health, Vol. II WOLZENDORFF: Armee-Krankeiten, in Realencyclopädie d. ges. Heilk. Bd. 1, p. 489 SCHUSTER: Kasernen, in VON PETTENKOFER UND ZIEMSSEN's Handbuch der Hygiene, II.Th. 2. Abth. CAMERON: Camps; Depopulation of Memphis; Epidemics of 1878 and 1879. Public Health, Vol. V, p. 152.]

CHAPTER XI.

MARINE HYGIENE.

THE melancholy accounts of the mortality from scurvy and typhus fever, which were formerly a part of the history of so many naval and passenger vessels, are happily now only records of the past. Occasionally, however, carelessness of the authorities, or of those responsible for the safety of people that 'go down to the sea in ships,' results in an outbreak of one or other of these diseases even at the present day. Thus, for the fiscal year ending June 30, 1882, seventy-one cases of scurvy and purpura were reported by the medical officers of the Marine Hospital service. It appears that in only one instance (where six cases of scurvy had occurred on one vessel) was any investigation ordered. A most superficial investigation showed that the law relating to the issue of lime-juice had been violated by the master of the vessel. No prosecution resulted. Such facts indicate that the laxness in the enforcement of the regulations expressly made to prevent this fatal disease, may be again followed by outbreaks of greater or less gravity.

THE SAILOR AND HIS HABITS.

Although the sailor of the present day, especially in the naval service, is morally and intellectually far in advance of the 'Jack Tar' of former days, his life, both afloat and ashore, leaves much to be desired on the score of temperance, chastity, or purity of thought and speech. The life of a sailor in the United States Navy only thirty years ago, is thus graphically described by Medical Director A. L. Gihon:* 'A

* Thirty Years of Sanitary Progress in the Navy. Annual address to the Naval Medical Society, Washington, 1884.

motley crew, of whom Americans were a minority, and
Englishmen, Irishmen, Northmen, and "Dagos" constituted the far greater part. Some had just returned
from another cruise, having squandered or been robbed
of their three years' pay by the landsharks, who
cajoled them, only half sober, to the rendezvous, to be
re-shipped, and thence to be herded, uncared for, on
the receiving ship, still popularly termed the "guardo,"
until drafted on board the first sea-going vessel. All
of them were in debt, most of them insufficiently clad,
and unable to properly outfit themselves. The wretched
herd, who were thus gathered from the purlieus of
Water street, and North street, and South street, who
at night were kennelled worse than dogs, by day fed
like them—crouching on their haunches around greasy
mess-cloths, cutting with jack-knives or pulling to
pieces with grimy fingers the chuncks of "salt horse"
and "duff" which made their daily fare, and which
later in the cruise were both spoiled and scanty,' did
not constitute an elevating subject for contemplation.

'Stint of good food,' continues Dr. Gihon, 'was,
however, not the chief of the old-time shell-backs'
ocean trials. Fed like a brute, housed worse than one,
however faithfully his labors were performed, there
was for him only a long, dreary season of imprisonment. For him there was no glad holiday on shore,
when the land broke the monotony of the waste of
waters. The officers might rush pell-mell out of the
ship, but Jack could only strain his longing eyes upon
the green fields or busy sea-ports. Notwithstanding
the hardships of the voyage, the wretched food, and
the outbreaks of disease, the crew were confined eight
months on board ship, before "general liberty" was
given, and then men and boys were sent on shore forty-
eight hours to indulge in a mad revel, and to return
crazed by rum, battered and bruised. The poor wretch

first made ravenously hungry for dissipation by his enforced confinement, was then expected to be temperate
in the feast of indulgence offered him, and punished
with vindictiveness if he sought to gorge himself with
the poor semblance of pleasure. The "cat" had been
abolished, but half a dozen boys strung upon the poop
"bucked and gagged" ; half a dozen men triced up by
their thumbs in the rigging ; each of the upright coffin-
like "sweat boxes" with its semi-asphyxiated inmate ;
the "brig" with its bruised and bloated crew in irons ;
the main-hold with its contingent under hatches ; the
sick-list swollen out of all proportion by inebriates,
injured men and venereal cases—these were the fruits
of the general liberty, which, within my professional
life represented the sum of sanitary interest in the
man before the mast.'

Under such circumstances, little could be hoped for
in the way of personal advancement of the crew. The
labors, however, of the writer just quoted, and others
among whom may be mentioned Wilson and Turner of
the navy, and Woodworth, Hebersmith and Wyman
of the Marine Hospital service, have drawn prominent
attention to the unsanitary conditions of the sailor's
life, and legal enactments have done much to elevate
him to his proper rank as a human being, entitled to
be treated with humanity at least.

The seaman in the navy now receives an abundance
of food, of good quality, usually well-cooked and
decently served. The sailor in the merchant service,
however, is still at the mercy of inhuman masters, who
exact excessive service in return for insufficient food,
abuse, ill-treatment and miserable lodging.

THE PASSENGER.

During the ten years from 1870 to 1879 inclusive,
passenger vessels carried 1,561,126 passengers from

foreign ports *en route* to New York city. The mean
duration of each voyage was 13.5 days. Out of the
above number of passengers 2,518 died on the voyage,
a death-rate of 1.01 per thousand for the voyage, and
43.5 per thousand per annum. These figures accentu-
ate the importance of sanitary improvement in passenger vessels. The causes of this excessive mortality
among emigrants, for it is almost exclusively among
the passengers in the steerage, or 'between decks,' that
the deaths occur, are over-crowding, improper feeding,
defective ventilation, filthy personal habits, and ineffi-
cient medical attention when sick. Although overcrowding is prohibited by statute, yet in every emigrant vessel that arrived in New York during the first
nine months of 1880, the number of passengers was in
excess of the number allowed by law.* The shorter
voyages and better sanitary conditions obtainable since
steamships, especially those built of iron, have come
into general use for the carriage of passengers, have
very much reduced the mortality on ocean voyages;
but as just shown, the death rate is still excessively
high, and many more improvements in the hygiene of
emigrant vessels and of passengers are desirable.

THE SHIP AS A HABITATION.

As a habitation for the sailor and passenger, the
ship demands the attention of the sanitarian. The
principal points in which he is interested are the construction and ventilation of sleeping apartments, and
the means of keeping the entire ship clean and free of
water and impure air.

The keel is the foundation of the ship. Branching
out transversely from it are curved timbers, the ribs,
which with the keel constitute the ship's frame. The
ribs are covered externally and internally with plank-

---

* TURNER: Hygiene of Emigrant Ships. Public Health, Vol. VI, p. 23.

ing, and the spaces between the two coverings are the
frame spaces, which are usually partly filled with filthy
water, decomposing organic matter and foul air. The
water collects in the bottom of the vessel, the bilge,*
whence it is pumped out of the vessel. If the pump
ing is neglected the bilge water becomes very offensive,
and may cause disease in persons exposed to exhalations from it. The frame spaces are rarely ventilated,
and hence are frequent sources of pollution of air on
board vessels.

The sleeping apartments of the crew of a merchant
vessel are in the forecastle, usually a dark, damp, un-
ventilated space in the bow of the vessel. On naval
vessels the crew sleep on the berth-deck, which in the
rarest instances is properly lighted and ventilated.
The berth-deck is usually below the water-line. In all
but the best class of vessels in the U. S. Navy, the air
allowance for each man is less than 100 cubic feet.

That a ship should above all be *seaworthy*, would
seem to require no argument. It is self-evident that a
leaky or rotten ship is at all times a highly dangerous
habitation; yet crews and passengers are almost daily
exposed to the perils of shipwreck in unseaworthy vessels, both in the mercantile marine and naval service.†

'Dampness, dirt, foul air, and darkness,' says
Gihon, 'are the direst enemies with which the sailor
has to battle when afloat.'‡ The first requisite for a
healthy ship is *dryness*. 'A damp ship is an unhealthy
ship,' says Fonssagrives, the greatest authority on naval hygiene. From official reports it appears that the
relative humidity of the berth-deck of vessels in the
U. S. Navy is nearly always above 80 per cent., very

---

* Hence called bilge water.

† Woodworth. The Safety of Ships and those who travel in them. Public
Health, Vol. III., p. 79 et. seq.

‡ Naval Hygiene. 3d Ed. p. 28.

often rising to 90 and 95 per cent.* From the same source it is learned that the class of respiratory diseases furnished, with one exception, the largest amount of sickness in the navy during the year 1880. It is the concurrent testimony of all authorities in marine hygiene that the vicious custom of daily drenching the decks with water, under the plea of cleanliness, is mainly responsible for this excessive moisture and its results. It is, therefore, one of the most important aims of marine hygiene to curtail this practice. Gihon recommends that the decks be coated with shellac to make them non-absorbent, and to wet them as rarely as possible, consistent with cleanliness.

The ship should be *clean and well ventilated*. Efforts to keep a ship clean should not be expended upon the decks only; the occupied apartments below the hatches, the bilges and frame-spaces should receive especial attention from the sanitary inspector. It is frequently necessary to remove the flooring of the vessels in order to expose the accumulations of filth, which often make an infected ship synonymous with a dirty ship. To disinfect a dirty ship, steam forced into the hold under pressure, after the filth has been cleaned out, gives the most satisfactory results. Sulphur and chlorine are next in efficiency. Solutions of sulphate of iron or chloride of zinc may be poured into the bilges to prevent decomposition.

It has been estimated† that a minimum of 400 cubic feet of air-space, with facilities for thorough ventilation, should be allowed to each person aboard ship. It is safe to say that no vessel that floats gives to her passengers or crew the advantages of such conditions. Ventilation of the holds and bilges, and of the spaces between the timbers or ribs, 'intercostal ventilation,'

---

* Report of Surgeon-General of the Navy, Washington, 1880.

Hygiene of Emigrant Ships, Public Health, Vol. VI., p. 26.

as Turner calls it, is especially necessary. Any system of ventilation that does not contemplate the removal of the foul bilge air, is unworthy of consideration by the sanitarian. The system briefly described in chapter I., p. 27, seems to fully meet the demands.

All parts of the vessel used as habitations or sleeping apartments should receive sufficient sunlight. At present, very few vessels have the quarters of the crew so disposed as to admit any sunlight at all.

In the fire-rooms of steamships, especially on that class of naval vessels termed monitors, the temperature often rises so high as would seem to render continued existence in it impossible. Gihon states that the average temperature in the fire-room of the monitor *Dictator*, is 145° F., while Turner states that in a vessel the average fire-room temperature was 167° F.* The stokers frequently suffer from heat-stroke, and in a very large proportion of cases from heart disease.

Lavatories and bathing facilities should be furnished on vessels for passengers and crew, and both should be compelled to keep their bodies and clothing clean.

DISEASES ON SHIPBOARD.

The diseases most liable to attack persons on shipboard are: Diseases of the respiratory organs, malarial diseases, digestive disorders, scurvy, typhus fever and skin diseases; and where the infection has been conveyed to the vessel by other persons, or by formites: yellow fever, cholera, small-pox and venereal diseases.

Most of these affections can be prevented by proper measures of hygiene, as demanded by the conditions described in this chapter, or by the enforcement of the following regulations:

Inspection of crews and passengers should be made compulsory before shipment. Persons suffering from

---

* Buck's Hygiene and Public Health, Vol. II., p. 190.

contagious or infectious diseases should not be taken on board.* In order to make this provision effective, the history of the individual for two weeks prior to his application for shipment should be known to the inspecting officer. Passengers should possess bills of health from the local authorities at their homes, in order that the presence or absence of such diseases as small-pox, yellow fever, cholera or plague may be established by the inspector. Cholera has always been introduced into this country by immigrants. Everybody admitted to the ship should be vaccinated. During several years past a number of epidemics of small-pox have been traced to foreign immigrants who had not been properly vaccinated.

Sailors should be submitted to a close personal inspection, and those suffering from venereal diseases should be rejected. The usual history of the cases is, that they soon go on the sick-list, and thus become an incumbrance instead of an aid on the vessel. These inspections should not be restricted to examinations for venereal diseases, but individuals incapacitated for the performance of a seaman's duties by any cause, should be rejected. This precaution would unquestionably reduce the number of marine disasters directly traceable to deficiency in the working force on board vessels. In this country the services of the medical officers of the Marine Hospital service, might be made available to carry out these inspections.

All sailors are liable to be placed in positions, where the prompt and accurate distinction of colors becomes necessary, hence all color-blind individuals should be rejected as seamen. The inability to distinguish colors has often been the cause of grave

---

* Gihon relates an instance where a man suffering from parotitis was transferred from the hospital of a receiving ship to a vessel going to sea. The disease was communicated to more than seventy of the crew of the latter vessel.

accidents at sea. Pilots can no longer obtain a license unless they satisfactorily pass an examination with reference to their ability to distinguish colors.

[The following works contain more detailed information upon the subject treated in the foregoing chapter:

A. L. GIHON: Practical Suggestions in Naval Hygiene, 3rd Ed., Washington, 1873. T. J. TURNER: Hygiene of the Naval and Merchant Marine; BUCK'S Hygiene and Public Health, Vol. II. WALTER WYMAN: Hygiene of Steamboats on the Western Rivers. Report of Supervising Surg. Gen'l, M: H. Service for 1882. Annual Reports of the Surgeon General of the Navy for 1879, 1880 and 1881. Various papers by J. M. WOODWORTH, A. L. GIHON, T. J. TURNER, HEBERSMITH, and A. N. BELL, in Public Health, Vols. I, III and VI.]

CHAPTER XII.

PRISON HYGIENE.

ALTHOUGH the frightful mortality which formerly seemed a necessary accompaniment of the life of the convict has in the past half century markedly diminished, the death rate among prisoners is still very greatly in excess of that of persons of the same age in a state of liberty.

The observations and labors of John Howard, the self-sacrificing philanthropist, in the latter half of the last century, and of Elizabeth Fry, in the first half of the present, directed the attention of legislators to the necessity of reform in the conduct of prisons and the treatment of prisoners. As a consequence of the labors of these reformers, the principles of prison discipline have been more fully developed during the past forty years by students of social science everywhere, and certain propositions have been formulated, which govern, to a greater or less degree, legislation upon this subject. These propositions are, briefly, as follows:

Prisoners must be properly classified, according to the nature of their crime and the duration of imprisonment.

The two sexes must be strictly separated, and no opportunity given for intermingling while in the prison.

Female prisoners should have female attendants exclusively. Male watchmen or other attendants should not be allowed in the female department of a prison.

All prisoners must be kept employed at some manual labor, not necessarily for profit, but as an agency in the moral reformation of the convict.

Punishments for infractions of discipline must not be excessive.

Efforts should be constantly made tending to the reclamation of criminals from their life of sin and crime.

Due care must be taken by the State to preserve the health and life of the prisoner whom the State has deprived of liberty, and the opportunity of taking care of himself.

A proper classification of prisoners, according to the degree of their criminality, the nature of the crime of which they have been convicted, or the length of time for which they have been sentenced, is now insisted upon by all students of prison discipline. As this subject more nearly concerns the social or legal relations of prisoners rather than their sanitary interests, it is here passed over with a mere mention.

The separation of the sexes, necessity of female attendants on prisoners of the same sex, employment of prisoners, and moral reformation of criminals likewise belong especially to the social aspects of the question, and can find no discussion in this place.

Regarding the remaining proposition, however, that which demands that the State shall exercise due care over the prisoner's health, it comprises a question that demands consideration in a text-book on hygiene.

There is now a general concurrence of opinion that the State, in depriving any person of liberty, has no right to subject the individual suffering such deprivation to any danger of disease or death. In other words, the State has no right to abbreviate the life of the convict sentenced to prison. This proposition requires that the State see to it that the prisoner is well-fed, well-clothed and well-housed; that he shall be well-cared for when sick, and that when his term of imprisonment expires, he shall be set at liberty, with only such effect upon his normal expectation of life as would result from the ordinary wear and tear of life upon his health.

It must be confessed, however, that the State is very far short of attaining this object. The mortality of convicts, even in the best regulated prisons, where especial attention is paid to the sanitary requirements of such buildings, is three times as great as among workmen in mines, confessedly one of the most dangerous occupations. If insurance companies desired to insure the lives of prisoners, the companies would be obliged, in order to secure themselves against loss, to make the premium equivalent to an advance in age of twenty years. This means that a free person has as long an expectation of life at forty years as a prisoner has at twenty. Attention is again called to the fact, that the conditions in the most favorably situated and liberally managed prisons only are here considered. What the results are in other institutions, less favorably constructed and managed, will be apparent from the following brief statement. Mr. George W. Cable has shown * that in some of the prisons in the southern States, under the vicious lease system, the mortality is eight to ten times greater than in properly constructed and managed prisons elsewhere. In Louisiana, for example, 14 per cent. of all the prisoners died in 1881; and in the convict wood-cutting camps of the State of Texas, one-half of the average number so employed during 1879 and 1880 died.

The mortality of prisoners is greatest in the second, third and fourth years of their confinement. In Millbank prison, in England, the death-rate per thousand was 3.05 in the first year, 35.64 in the second, 52.26 in the third, 57.13 in the fourth, and 44.17 in the fifth years of imprisonment.

The diseases most frequent among prisoners are pulmonary phthisis, and diseases of inanition, manifested by general dropsy. Consumption furnishes from

---

* Century Magazine, February, 1884.

40 to 80 per cent. of all deaths. When prisoners are attacked by acute febrile, or epidemic diseases (small-pox, cholera, dysentery), the mortality is much higher than among persons in a state of liberty. This fatality is due to an anémic or cachectic condition, which has been called 'the prison cachexia,' a depravement of constitution which yields readily to the invasion of acute diseases.

Prisons should be built upon a healthy site, be properly heated and ventilated, have an abundant water supply, and facilities for a prompt and thorough removal of sewage. Baths and lavatories should be conveniently arranged, in order that thorough cleanliness can be enforced.

The problem of feeding prisoners requires careful study. The food should not merely be sufficient in quantity and of good quality, but it should be well-cooked, and the bill of fare varied often, in order to avoid creating a disgust by an everlasting sameness. Prisoners often suffer from nausea and other digestive derangements, brought on solely by the monotonous character of the daily food.

In work-shops and sleeping-rooms, dormitories or cells, the cubic air-space allowed to each inmate should not be less than 600 cubic feet, with proper provision for ventilation. The use of dark or damp cells as places of confinement is a relic of the barbarism in the treatment of convicts against which John Howard raised his voice so effectively in the last century. An abundance of sunlight should be admitted into every room in which a human being is confined.

An important hygienic measure is daily exercise in the open air. It should be regularly enforced, and its modes frequently varied in order that it may not degenerate into a mere perfunctory performance.

Punishment for infractions of the prison-discipline should be inflicted without manifestation of passion, and only under the immediate direction of some official responsible to the State. It is questionable whether physical punishments, such as whipping, tricing up by the thumbs with the toes just touching the floor, bucking and gagging, and similar barbarities should be permitted under any condition. The permission to exercise such power is extremely liable to be abused by officials. The system of leasing out prisoners to private parties, which prevails in some of the southern United States is vicious in the extreme, because it places the convict under the control of persons not responsible to the State, and in the majority of instances, morally unfitted to wield the power of inflicting punishment.

[The following works on Prison Hygiene and Prison Reform are recommended to the student:

A. BAER: Gefängniss-Hygiene, in VON PETTENKOFER UND ZIEMSSEN'S Handbuch der Hygiene, II. Th., 2 Abth. Trans. International Penitentiary, Congress, London, 1882. Trans. National Prison Association, Baltimore, 1872. G. W. CABLE: The Convict Lease System in the Southern States, Century Magazine, February, 1884.]

CHAPTER XIII.

EXERCISE AND TRAINING.

The healthy functions of the bodily organs can only be maintained by more or less constant use. A muscle, or other organ that is unused soon wastes away or becomes valueless to its possessor. On the other hand, trained use of the various organs makes them more effective for the performance of their functions. Thus, by practice, the eye can be trained to sharper vision, the ear to distinguish slight shades of sound, the voice to express varying emotions, the tactile sense to accurately appreciate the most minute variations of surface and temperature, and the hand to greater steadiness, or the performance of difficult and complex feats. The effectiveness of other organs, muscles, or groups of muscles can also be increased by systematic training, as.is seen in the athlete and gymnast.

PHYSIOLOGICAL EFFECTS OF EXERCISE.

When a muscle contracts, the flow of blood through it is increased. Hence, contraction of a muscle, which consumes or converts stored-up energy, at the same time draws upon the circulation for a new supply of food-material to replace that consumed. The activity of the circulation through a muscle in action, results in increased nutrition and growth of the muscle.

During muscular action, the activity of the respiratory process is increased. A larger quantity of air is taken into the lungs, more oxygen is absorbed by the blood, and an increased elimination of carbonic acid, takes place. The experiments of Pettenkofer and Voit show that while in a state of rest, the average absorption of oxygen in twelve hours amounted to 5,771.56

grains, during work the amount reached 8,410.44 grains. For the same period the elimination of carbonic acid was: during rest, 8,825.25 grains, during work, 13,217.50.

Upon the circulation, muscular exercise likewise exerts a manifest influence. The action of the heart is increased both in force and frequency, the arteries dilate, and the blood is sent coursing through the system more rapidly than when the body is at rest.

Cutaneous transpiration is also promoted by muscular exercise. It is probable that in this way some of the effete matters in the system are removed, being held in solution and carried through the skin in the perspiration.

PHYSICAL TRAINING.

There can be no question that systematic training of the muscles has a favorable influence upon health and longevity. Persons who are actively engaged in physical labor, other things being equal, are healthier, happier, and live longer than those whose occupation makes slight demands upon their muscular system. In default of an active occupation the latter class is forced, if good health is desired, to adopt some form of exercise which will call the muscles into activity.

The principal methods of physical training are, walking or running, rowing, swimming, and the various in-door gymnastic exercises. Rapid walking or running is one of the best methods of physical exercise, for not only are the muscles of the legs and thighs developed, but the capacity of the chest is increased— one of the principal objects of physical training. By combining walking with some form of in-door gymnastics, such as practice with the dumb-bells, Indian clubs, rowing machine, or pulley weights, nearly all the good effects of the most elaborate system of training can be obtained.

For the gymnastic exercises, various forms of useful labor may be substituted with advantage, such as wood-chopping or sawing, or moderate work at any physical labor.

The scheme of studies in our public school system should include physical training for both sexes. This is a question not merely of individual, but of national importance. Weak and unhealthy children are not likely to grow up into strong and healthy men and women ; and the latter are necessary for the perpetuity of the nation. The time seems to have arrived when physical education should no longer be looked upon as a whim of unpractical enthusiasts and hobby-riders, but as an indispensible element in every school curriculum.

There is a tendency among instructors in physical training to make their systems too complicated, or dependent upon expensive or cumbersome apparatus. This is to be deprecated. All the muscles of the body can be called into action by very simple exercises, easily learned, and readily carried out.

An important preliminary to all methods of training, is a thorough physical examination of the pupil by a competent physician, in order to determine whether certain exercises are allowable. For example, in all organic heart affections, exercises of a violent character must be interdicted. A boy or man with valvular disease of the heart, cannot run, row or swim with safety. The organ is easily overtasked in this condition and liable to fail in its function.

One of the simplest and best methods to cause the pupil to assume a correct position of the body, and to acquire ease and grace in his movements, is to teach him the 'setting up,' as practiced in the U. S. army.*

In walking, a free, swinging step should be acquired, with the head erect, shoulders thrown back

---

* UPTON's Infantry Tactics. School of the Soldier, lesson I.

and chest well to the front, the whole body from the hips upward inclining slightly forward. The clothing should be loose around the upper part of the body, in order not to interfere with the freest expansion of the chest, and to give the lungs and heart ample room for movement. Even in-door gymnastic exercises alone, when practiced under intelligent supervision, will accomplish very favorable results, as shown by the following table:

TABLE I.

Showing average state of development on admission to gymnasium; average state of growth and development after six months' practicing two hours a week, and average increase during that time. (Bowdoin College Gymnasium, under Dr. D. A. Sargent. 200 students from the classes of 1873–77, *inclusive. Average age,* 18.3 *years.*)*

	ON ADMISSION.	AFTER 6 MONTHS' PRACTICE.	AVERAGE INCREASE.
Height, - -	5 ft. 8 in.	5 ft. 8 ¼ in.	¼ in.
Weight, -	135 lbs.	137 lbs.	2 lbs.
Chest (inflated),	35 in.	36¾ in.	1¾ in.
Chest (contracted,)	32¼ in.	33 in.	¾ in.
Forearm, -	10 in.	10¾ in.	¾ in.
Upper arm (flexed,)	11 in.	12 in.	1 in.
Shoulders (width),	15½ in.	16¼ in.	¾ in.
Hips, - -	31½ in.	33¾ in.	2¼ in.
Thigh, -	19½ in.	21 in.	1½ in.
Calf, - -	12½ in.	13¼ in.	¾ in.

OVER-EXERTION.

However necessary for the preservation of health physical exercise may be, over-exertion should be carefully avoided. Over-strain and hypertrophy of the heart are often results of excessive exertion. Dr. Da Costa has described a form of 'irritable' and weak heart occurring especially among soldiers, which he has clearly traced to over-exertion. Severe labor and violent athletic exercises have been followed by like serious results. Long distance pedestrianism has furnished within recent years quite a number of individ-

* Apparatus used: Weights, 10-15 lbs.; Dumb-bells, 2½ lbs.; Indian Clubs, 3½ lbs.; Pulleys.

uals who were broken down in health by the excessive strain on the physical organisation involved. Cardiac strain is not infrequent among this class.

Spasm, paralysis, or atrophy of muscles sometimes results, when these are exhausted by uninterrupted or excessive exercise. This effect is shown by writer's and telegrapher's cramp, and similar affections. For these reasons it is important that both exercise for health and actual work should be so regulated as to conduce to the individual's benefit, and not to his detriment.

[On the subjects embraced in this chapter the following works may be studied with advantage :

A. BRAYTON BALL: Physical Exercise, in BUCK's Hygiene and Public Health, Vol. I   WM. BLAIKIE: How to Get Strong and How to Stay So. A. MACLAREN: Training in Theory and Practice.]

CHAPTER XIV.

BATHS AND BATHING.

THE most important sanitary object of bathing is cleanliness. A secondary object of the bath is to stimulate the functions of the skin, and to produce a general feeling of exhilaration of the body. Baths are used of various temperatures. A cold bath has a temperature of from 40° to 75° (Fahr.); a tepid bath from 75° to 85°; a warm bath from 85° to 100°; and a hot bath from 100° to 110.°

Tepid, warm, or hot baths are used principally as cleansing agents, or as therapeutic measures. They cause dilatation of the cutaneous capillaries, diminish blood-pressure, and reduce nervous excitability. The hot bath is also a method for restoring warmth to the body in certain cases of shock, or to remove the immediate effects of injurious exposure to low temperature.

The so-called Russian and Turkish baths, so popular in the larger cities of this country, are modifications of vapor and hot-air baths, or rather combinations of these with cold baths. The Turkish bath is especially to be recommended for its depurative and invigorating effects.

Cold baths are used not merely for their cleansing effects, but principally for their stimulating effects upon the system. When first plunging into a cold bath, there is usually a momentary shock, the respiration is gasping, and the pulse is increased in frequency. These symptoms disappear in a few moments however, and reaction follows. To a healthy person a cold·bath is a delightful general stimulant, removing the sense of fatigue after physical exertion, and causing an extremely refreshing sensation throughout the body.

As a therapeutic measure, the cold bath has a wide field of usefulness. For the reduction of the bodily temperature in fevers, and inflammatory diseases, and especially in heat-stroke, it is more prompt and effective than any other agent at the command of the physician.

Sea Bathing.—The most stimulating form of the cold bath is doubtless the salt-water bath as taken at the sea-shore. The revulsive effect of the impact of the waves and breakers upon the skin, and the stimu-lation due to the saline constituents of the sea-water heighten the invigorating effects of the simple cold bath. The beneficial results of sea-bathing are, how-ever, not entirely due to the bath, but are to a great degree dependent upon the bracing air of the sea-shore, absence of the care and anxieties of business, and the temporary change in food and habits that a residence at the sea-side involves. Nevertheless, salt-water baths are more stimulant to the skin than those of simple water, and part of the good effects of sea-bath-ing can often be obtained from a salt-water bath taken at home. The following mixture of salts dissolved in about thirty gallons of water for one bath, makes a fairly good substitute for a sea-bath:

Take of Chloride of Sodium (common salt),   -   9 lbs.
    Sulphate of Sodium (Glauber's salt),   -   4 "
    Chloride of Calcium,   -   -   -   -   ¼ "
    Chloride of Magnesium,   -   -   -   - 3½ "

There is a prevalent popular belief that it is extremely dangerous to enter a cold bath when heated or perspiring. The author is of opinion that this belief is erroneous. The stimulant and bracing effects of the cold bath are most manifest if it be taken while the individual is very warm or bathed in perspiration. Several years ago the author made a series of observa-tions upon himself to determine the effects of the cold bath when the body was very warm. Every afternoon a free perspiration was provoked by a brisk walk of about a mile and a half in the sun. As soon as the clothing could be cast off, and while the body was still freely perspiring, a plunge was taken into a fresh-water bath of about 60° Fahr. No ill results followed; on the contrary, the sensation immediately following the bath, and for six or eight hours afterward was exceedingly pleasant. The health remained perfect, and the weight decidedly increased during the two months the practice was continued. There is probably no danger to a healthy person in this practice, but it is considered advisable to immerse the head first ('take a header'), to avoid increasing the blood-pressure in the brain too greatly, which might result if the body were gradually immersed from the feet upward.

RULES FOR BATHING.

The following series of rules have been issued by the English Royal Humane Society, and are well worth observing by bathers:—'Avoid bathing within two hours after a meal. Avoid bathing when exhausted by fatigue, or from any other cause. Avoid bathing when the body is cooling after perspiration. Avoid bathing altogether in the open air, if, after having been a short time in the water, there is a sense of chilliness, with numbness of the hands and feet; but bathe when the body is warm, provided no time is lost in getting into the water. Avoid chilling the body by sitting or standing undressed on the banks, or in boats, after having been in the water. Avoid remaining too long in the water, but leave the water immediately if there is the slightest feeling of chilliness. The vigorous and strong may bathe early in the morning on an empty stomach. The young and those who are weak had better bathe two or three hours after a meal; the best time for such is from two to three hours after breakfast.

Those who are subject to giddiness or faintness, or suffer from palpitation or other sense of discomfort at the heart, should not bathe without first consulting their medical adviser.'

To these instructions may properly be added that a warm or hot bath should be avoided, if the person is liable to exposure to cold within a few hours after the bath; that women should, as a rule, not take a cold bath while menstruating, or during the last two months of pregnancy, and that persons suffering from organic heart disease should especially avoid surf-bathing.

After bathing, the body should be thoroughly dried with soft towels, otherwise eczematous eruptions are liable to follow in the parts subject to friction from opposing surfaces of the skin, as in the groins, the perineum and inner surface of the thighs, the armpits, or the under surface of the breasts, in women in whom these organs are large and pendant.

Friction of the skin with a coarse towel, or so-called 'flesh brush' is a popular practice, but is not to be universally commended. The hyperemia of the surface thus produced may sometimes induce cutaneous diseases (erythema, eczema, psoriasis,) in those predisposed.

DANGERS OF COLD BATHING.

One of the most serious dangers of cold bathing, but which is not sufficiently appreciated, is the tendency to nausea and vomiting if the stomach contains much food. There can be no doubt that many of the cases that are called 'cramp,' and which frequently result in drowning, are due to this cause.*

Cramps of various muscles sometimes occur, rendering the bather helpless, and if in deep water he is liable to drowning before assistance can reach him.

---

* So far as the author is aware, Dr. John Morris, of Baltimore, first called especial attention to this source of danger.

HOW TO RESTORE THE APPARENTLY DROWNED.

In drowning, death takes place by asphyxia. The respiration is arrested by the submersion of the head, the carbonised blood gradually poisons the system, and the heart ceases to beat. So long as the heart will react to its appropriate stimulus, the person may be restored to life. The first thing to do therefore after a recently drowned person is taken out of the water, is to attempt to re-establish the arrested respiration. Several methods are in use for this purpose. Sylvester's is one of the simplest. It is as follows:

The body being placed on the back (either on a flat surface, or better, on a plane inclined a little from the feet upward), a firm cushion, or similar support (a coat rolled up will answer) should be placed under the shoulders, the head being kept in a line with the trunk. The tongue should be drawn forward to raise the epiglottis and uncover the wind-pipe. The arms should be grasped just above the elbows and drawn upward until they nearly meet above the head, and then at once lowered and replaced at the side. This should be immediately followed by pressure with both hands upon the belly, just below the breast-bone. The process is to be repeated fifteen to eighteen times a minute.

Several years since the Michigan State Board of Health published a method which is comprehensive, effective, easily understood and readily carried out. This method has also been adopted by the U. S. Life Saving Service. The following are the details of the 'Michigan method':

RULE 1.—REMOVE ALL OBSTRUCTIONS TO BREATH—ING. *Instantly* loosen or cut apart all neck and waist bands; turn the patient on his face, with the head down hill; stand astride the hips with your face toward his head, and locking your fingers together under his belly, raise the body as high as you can without lifting the forehead off the ground, and give the body a smart jerk to remove mucus from the throat and water from the windpipe, hold the body suspended long enough to slowly count *one--two--three--four--five*, repeating the jerk more gently two or three times.

RULE 2.—Place the patient on the ground face downward, and, maintaining all the while your position astride the body, grasp the points of the shoulders by the clothing, or if the body is naked, thrust your fingers into the armpits, clasping your thumbs over the points of shoulders, and *raise the chest as high as you can* without lifting the head quite off the ground, and hold it long enough to slowly count *one—two—three*. Replace him on the ground with his forehead on his flexed arm, the neck straightened out, and the mouth and nose free ; place your elbows against [the inner surface of] your knees and your hands upon the sides of his chest *over the lower ribs and press downward and inward* with increasing force long enough to slowly count *one—two*. Then suddenly let go, grasp the shoulders as before, and raise the chest ; then press upon the ribs, etc. These alternate movements should be repeated ten or fifteen times a minute for an hour at least unless breathing is restored sooner. Use the same regularity as in natural breathing.

RULE 3.—After breathing has commenced, *restore the animal heat.* Wrap him in warm blankets, apply bottles of hot water, hot bricks, or anything to restore heat. Warm the head nearly as fast as the body, lest convulsions come on. Rubbing the body with warm cloths or the hands and slapping the fleshy parts may assist to restore warmth and the breathing also.

If the patient can *surely* swallow, give hot coffee, tea, milk, or a little hot sling. Give spirits sparingly, lest they produce depression.

Place the patient in a warm bed, and give him
plenty of fresh air. Keep him quiet.

BEWARE! *Avoid delay.* A *moment* may turn the
scale for life or death. Dry ground, shelter, warmth,
stimulants, etc., at this moment are nothing—*artificial
breathing is everything*—is the *one remedy*—all others
are secondary. *Do not stop to remove wet clothing.*
Precious time is wasted, and the patient may be fatally
chilled by exposure of the naked body, even in summer.
Give all your attention and efforts to restore breathing
by forcing air into, and out of, the lungs. If the
breathing has just ceased, a smart slap on the face, or
a vigorous twist of the hair will sometimes start it
again, and may be tried incidentally. Before natural
breathing is fully restored, do not let the patient lie on
his back unless some person holds the tongue forward.
The tongue by falling back may close the windpipe and
cause fatal choking.

Do not give up too soon; you are working for life.
Any time within two hours you may be on the very
threshold of success without there being any sign
of it.*

PUBLIC BATHS.

In all large cities and towns, provision should be
made for free public baths, conducted under official
supervision, and for the especial use and benefit of the
poorer classes. General cleanliness is not merely a
factor in the preservation of the public health, but
there is good reason to believe that the cause of good
order and decency would likewise be promoted, by furnishing the public the means of easily and cheaply
keeping clean. Several of the larger cities in the
country have established public baths upon a limited

---

* Report Michigan State Board of Health, 1874, p. 91-99.

scale, but it is reported that in one of them—Philadelphia—they have been discontinued from economical
motives.

[Many valuab'c hints on sea-bathing can be obtained from the little work
on Sea Air and Sea-Bathing, by Dr. Jno. H. Packard, published in the
series of *American Health Primers*.]

CHAPTER XV.

CLOTHING.

THE primary object of clothing is the protection of the body against injurious influences of heat, cold and moisture. Secondarily, the moral sense of civilised communities demands that the nude human body shall not be exposed in public. Hence, there are moral as well as sanitary reasons for the wearing of clothing; only the latter can be considered in this place.

Bodies radiate or absorb heat, accordingly as they are surrounded by a medium having a lower or higher temperature than themselves. In order, therefore, to avoid chilling of the human body, if exposed to a temperature below 98° Fahr., clothing must be worn to prevent or retard radiation of the body-heat. Exposure of the unprotected body to a low temperature would not merely cause chilling of the surface, owing to the rapid loss of heat, but would incidentally produce congestion of internal organs, by causing constriction of the superficial capillaries.

Clothing is also worn as a protection against great heat. The head, especially, needs protection from the sun's rays.

CLOTHING MATERIALS.

The materials from which clothing is made are, principally, cotton, linen, wool, silk, and the skins of animals. Of these, probably the most universally used is *cotton*. It is cheap, durable, does not shrink when wet, absorbs little water, and conducts heat readily. It is, therefore, especially valuable for summer garments, allowing rapid dissipation of the body heat and evaporation of the perspiration.

Linen conducts heat even better than cotton, and is for this reason largely used for summer clothing. Its principal advantage over cotton is, that it is more durable, and less harsh to the skin.

Wool absorbs water readily and is a bad conductor of heat. It is, therefore, valuable as a winter garment, retarding radiation from the body. Woolen undergarments should be worn at all seasons, in order to prevent too rapid changes of the surface, and so invoking diseases depending upon chilling of the body. Clothing of pure wool (flannels), are liable to irritate the skin of some persons. A mixture of wool and cotton, known as 'Saxony wool,' is softer and less irritating, and makes a serviceable substitute for pure wool.

Silk is often used for undergarments. It is light, soft, and a bad conductor of heat.

The skins of animals, with the fur on, are often used for outside clothing. They furnish great protection against severe cold. The skin is impermeable to wind and rain, while the thick, pilous covering of fur retards to a very great degree the radiation of heat. In British America, the northwestern States and Territories and in the arctic regions, the use of skin clothing is necessary for comfort.

As a protection against moisture (rain and snow), *rubber cloth* is used for overcoats, etc. While it serves effectually in keeping out the rain, it prevents evaporation of the perspiration, increasing the liability to chill, and rendering the person wearing it very uncomfortable, except in cold weather.

Leather is used almost exclusively in the manufacture of foot-wear. It is sometimes used, however, for other articles of clothing, such as coats, trowsers, etc. It furnishes most effective protection against cold.

The *color* of the clothing is of great importance. Exposed to the sun, white wool or silk absorbs very little more heat than linen or cotton, but the same material, of different colors, when exposed to the sun's rays, exhibits marked differences in absorptive capacity. The following table shows the results of some experiments of Pettenkofer. The material used was cotton shirting of the colors named :

White absorbed - -	100	heat units.
Light Sulphur Yellow absorbed	102	"
Dark Yellow absorbed -	140	"
Light Green absorbed -	155	"
Turkey Red absorbed -	165	"
Dark Green absorbed -	168	"
Light Blue absorbed -	198	"
Black absorbed - -	208	"

When protected from the sun's rays, however, the material becomes important and the color is of little consequence. Wool, being a bad conductor of heat, retards radiation from the body, and is hence the best material for winter clothing.

Gases and vapors, probably also disease-germs, are absorbed by clothing and may be thus conveyed from place to place. It has been found that woolen clothing possesses this power of absorption to a much greater degree than linen or cotton. The bad odor of a crowded room or of tobacco smoke frequently clings to woolen garments for days, although they may be exposed constantly to the air during the interval. It would be advisable, therefore, that physicians attending infectious diseases, hospital attendants and nurses, should wear linen or cotton clothing instead of woolen.

Clothing should be made to *fit* properly. It should not restrain muscular movements, obstruct the circulation or compress organs. Hence, corsets, belts and garters are to be condemned. It is a fact of common observation, that moderately loose clothing is warmer than close-fitting.

Especial attention should be given to the shape and fitting of foot-wear. Boots and shoes are usually made with little regard to the physiological anatomy of the foot, and as a result the feet of most Americans are deformed, beauty and usefulness being in a great degree sacrificed to the Moloch of fashion.*

Dyes used for coloring fabrics are sometimes poisonous. The author has repeatedly seen troublesome eruptions and even ulcerations of the legs from wearing stockings dyed with aniline compounds.

By appropriate treatment clothing can be made *non-inflammable*. Tungstate and phosphate of soda are used to reduce the inflammability of fabrics. The addition of 20 per cent. of tungstate of soda and 3 per cent. of phosphate of soda to the starch sizing used for stiffening linen is effective. The material is not injured by it, and a smooth surface and polish can be obtained under the hot iron. Prof. Kedzie has recommended borax for the same purpose. He says: 'The simplest and easiest way to make your cotton and linen fabrics safe from taking fire is to dissolve a heaped teaspoonful of powdered borax in half a pint of starch solution. It does not injure the fabric, imparts no disagreeable odor, and interferes in no way with the subsequent washing of the goods. It does not prevent the formation of a smooth and polished surface in the process of ironing. Borax can be found in every village, and is within the reach of all. It is a cheap salt, and its use for this purpose is very simple.'†

[The following works may also be studied to advantage :

HAMMOND: Hygiene, p. 579. L. MEYER : Kleidung, in Realencyclopædie d. ges. Heilk., Bd. VII, p. 446. VAN HARLINGEN : Care of the Person, in BUCK's Hygiene and Public Health, Vol. I.]

---

* See a practical paper by DR. BENJ. LEE : A Shoe That Will Not Pinch. Sanitarian, June, 1884, p. 493.

† Michigan State Board of Health, 1880, p, 181.

CHAPTER XVI.

DISPOSAL OF THE DEAD.

WHEN life is extinct in the animal body, decomposition begins. This may be either putrefactive or non-putrefactive. The difference between the two processes has been explained by Leibig. In putrefaction of organic matters, only the elements of water take part in the formation of the new compounds which result, while in non-putrefactive decomposition or decay, the oxygen of the air plays an important part. Putrefaction can go on under water, while decay can only take place when the supply of free oxygen is abundant.

The prompt removal of the bodies of the dead from the immediate vicinity of the living is a matter of prime sanitary importance. If death results from a contagious or an infectious disease, the necessity for the removal of the corpse is evident. But, even where there is no danger of propagation of infectious disease, the products of putrefaction and decay may give rise to serious derangements of health if allowed to pollute the air.

The chief methods of disposal of the dead are burial in the earth, entombment in vaults, and cremation.

INTERMENT.

The most common method of sepulture is burial in the earth. The corpse is usually enclosed in a case (coffin) of wood, or metal, and buried from four to six feet deep. Here decomposition sets in, which is at first putrefactive, and later on non-putrefactive. In the course of several years, from five to ten, the entire body with the exception of the bones has usually disappeared, and becomes converted into a dry mould.

The soil of a burial ground should be dry and porous, so as to be easily permeated by the air. In a sandy or gravelly soil, the decay of a corpse is much more rapid than in a moist, clayey soil. In the latter, the bodies more readily undergo putrefaction or become converted into a substance termed adipocere. It has been calculated that in a gravelly soil the decay of a corpse advances as much in one year, as it would in sand in one and two-thirds, and in clay in two to two and one-third years. The decay of the dead bodies is principally (if not entirely) dependent upon the presence of living vegetable organisms. If the access of free oxygen is prevented, the bacteria of putrefaction will thrive and cause putridity. If, however, the soil is loose, porous, and easily permeable by the air, the bacteria of decay will be present and produce their characteristic effects.

The barometric pressure seems to affect the decomposition of dead bodies. For example, at the refuge of St. Bernard in the high Alps, the bodies of those dying are not buried, but exposed to the air, where they undergo a drying, shrinking and mummification, instead of putrefaction or decay.

Alternate saturation and drying of the soil promotes the rapidity of decay.

Certain occupations are said to produce changes in the tissues which resist decay. Thus tanners are supposed to resist the final changes of the tissues longer than persons of other occupations. Shakespeare makes the grave-digger in Hamlet say: 'A tanner will last you nine years.' The corpses of those poisoned by phosphorus, arsenic, sulphuric acid or corrosive sublimate also decay more slowly than those of cases of infectious diseases.

All the tissues may be converted into adipocere, but in the large majority of cases only the fat and connective tissue undergo this change.

SUPPOSED DANGERS OF BURIAL GROUNDS.

Popular sanitary literature teems with supposed instances of the injurious influences of cemeteries upon the health of persons living in their vicinity. An unprejudiced consideration of the subject shows, however, that there is no trustworthy evidence that any of the gases exhaled by decaying or putrefying bodies are injurious to health. The air of closed burial vaults may be dangerous from the large proportion of carbonic acid contained in it, but the other gaseous products of decomposition have no deleterious effects. The dangers to health from the proximity of cemeteries are doubtless very much exaggerated. Pettenkofer and Erismann have shown that a single large privy vault, containing about 600 cubic feet of excrement, gives off nearly as large an amount of putrefactive gases in the course of one year, as is exhaled by a burial ground containing 556 decomposing corpses in ten years.

Where bodies are properly buried, and the ground is not overcharged by corpses, it is not probable that infectious diseases are propagated from interred bodies. There are no facts on record that show that such an event has occurred.

The dangers of pollution of water by cemeteries have also been much overestimated. The purifying power of soil strata through which the water is compelled to percolate before reaching the well, after becoming charged with the products of decomposition, is in most cases sufficient to remove all deleterious matters.

Cemeteries should not be located within a city, but must be easily accessible. The soil should be dry gravel or sand, with a low ground-water level. The graves need not be deeper than four feet to the top of the coffin.

ENTOMBMENT IN VAULTS.

Burial vaults in churches, or in the open air should be discountenanced. The gases of decomposition are given off directly to the air without the modifying power of the soil, and often constitute a nuisance, even if not deleterious to health. Entombment in vaults or crypts has not a single favorable circumstance to recommend it.

CREMATION.

Within recent years the rapid incineration of the dead in properly constructed furnaces has been frequently recommended. In the United States a cremation furnace was built several years ago at Washington, Pa., by the late Dr. J. C. LeMoine. Among the remains of those cremated were those of the late Dr. Samuel D. Gross, the distinguished surgeon. The practice has not gained very many adherents, however, although cremation societies have been organised in several of the cities throughout the country. Aside from the objections urged by the more conservative classes who desire to adhere to the time-honored custom of interment, serious legal objections have been brought forward. In cases where poisoning was suspected some time after death, the cremation furnace would have destroyed every evidence of crime, and conviction of a criminal poisoner could not be obtained.

The real advantages of cremation, such as rapid destruction of a corpse, economy of space in keeping the remains, and avoidance of pollution of the soil by decaying bodies, and possible pollution of air and water, are more than counter-balanced by the expense and the medico-legal objection mentioned. From a sanitary point of view, cremation is not necessary in this country. A proper regulation of cemeteries will prevent any possible dangers to the living from pollution of the air, soil or water by the decaying remains of human beings.

INTERMENT ON THE BATTLE FIELD.

After battles, the disposal of the bodies of the slain is often a serious problem. Nægeli proposes the following method of interment: After selecting the place of burial, the sod and layer of humus are removed from a sufficiently large surface and thrown to one side. The corpses are then laid upon the denuded place, and the layers of corpses separated by sand, gravel, or fine brush-wood. A trench is then dug around the pile of dead and the soil gained is thrown over the corpses until they are covered to the depth of three feet, when the humus and sod are placed over the whole. This furnishes a dry grave in which decay rapidly takes the place of putrefaction, and the corpses soon moulder away. The same procedure may be followed in cases of epidemics where the number of deaths is too great to properly bury them in single graves.

CHAPTER XVII.

THE GERM THEORY OF DISEASE.

THE ruling doctrine in the pathology of the present day is the germ theory of disease. Based upon the doctrine of *omne vivum ex vivo*, and supported by strong experimental and clinical evidence it is accepted by the great majority of physicians. Its advocates claim that the large class of diseases known as contagious or infectious, are all due to the presence in the blood or tissues, of minute organisms, either animal or vegetable. Many other diseases, not at present included in the above class by general pathologists, are also believed, by the adherents of the germ theory, to be caused in the same way. The following constitutes a brief review of the most prominent facts in the history of the doctrine:

The doctrine of the vital nature of the contagium of disease—the *contagium animatum* of the older writers—was held in a vague way by many of the physicians of the past, but it was not until the latter part of the last century that the theory took definite shape. In the works of Hufeland, Kircher and Linné, the idea is expressed with more or less directness that the propagation of infectious diseases depends upon the implantation of minute independent organisms into or upon the affected individual. This hypothesis was, however, first clearly enunciated, and defended with great force, by Henle, in 1840. Three years earlier, Cagniard de la Tour and Schwann, had established a rational basis for the theory by their observations upon the yeast plant and its relation to fermentation. In 1835, Bassi had discovered in the bodies of silkworms affected by *muscardine*, a disease of these insects which proved very destructive, a parasite which was soon shown to be the cause of the disease. Within the next few years, Tulasné, DeBary and Kuehn, proved that certain fungi were the causes of the potato rot and other diseases of plants. Schœnlein, Malmsten and Gruby, between 1840 and 1845, demonstrated that those skin diseases of man classed as *the tineœ*, were due entirely to the action of vegetable parasitic organisms.

Up to this time the germ theory, as now accepted, had received no support from experiment. All the diseases claimed as parasitic were purely local; so far as the parasitic nature of the general diseases was concerned, all was hypothetical. In 1849, Guérin Méneville discovered a corpuscular organism in the blood of silkworms affected by the *pebrine*, which was later proven by Pasteur to be the true cause of this destructive disease. Pollender, in 1855, and Brauell, in 1857, found numerous minute rod-like organisms (bacteria) in the blood of animals dead with splenic fever. In 1863, Davaine, investigated the subject more fully, and showed beyond doubt that the little organisms discovered by Pollender were the true cause of splenic fever, or anthrax. The more recent researches of Koch upon the history of these bacteria or bacilli of splenic fever have removed all doubt of their etiological significance.

The careful observations and researches of such scientific investigators as Rindfleisch, Waldeyer, Von Recklinghausen, Chauveau, Billroth, Carter, Burdon Sanderson, Koch, and others, have established the theory upon a secure foundation. There is an increasing number of diseases for which the parasitic origin may be accepted as fully proven. Among these are splenic fever, relapsing fever, erysipelas, actinomycosis, glanders and diphtheria. For another class of diseases including tuberculosis, pleuro-pneumonia of cattle,

cholera, typhoid fever and croupous pneumonia, the
etiological connexion between the organisms, and the
disease, appears probable. For a very large remaining
class of diseases, however, there is no trustworthy
evidence to show their origin from parasitic micro-
organisms.

In connexion with the germ theory, there has
arisen of late a very important question in its bearing
upon preventive medicine. This is the value of the
so-called protective inoculations against infectious
diseases. The protective influence of vaccination
against small-pox is firmly established by indubitable
evidence. Within the last three or four years a procedure introduced by Pasteur to protect animals against
certain fatal infectious diseases, such as splenic fever,
fowl cholera and rabies, has claimed much attention.
Pasteur's observations were first made upon the disease
termed chicken cholera. He found that the blood of
the dead fowls, or of those attacked by the disease
swarmed with bacteria. Inoculation of healthy fowls
with this diseased blood, or with the bacteria alone,
carefully freed from all animal fluids, produced the
same disease. The bacteria were therefore assumed to
be the cause of the disease. The investigator then took
a quantity of these bacteria and 'cultivated' them
through a number of generations, using sterilised
chicken broth as a culture medium. Fowls inoculated
with the result of the last cultivation were still
attacked by the same symptoms but in a very mild
degree, and almost uniformly recovered from the
disease. On subsequent inoculation with infected
blood, no effect was produced upon the 'vaccinated'
fowls, while the same blood introduced into fowls not
'protected' by the previous inoculation, produced its
customary fatal effect. Pasteur and others repeated
these experiments with the organisms found in the

blood in splenic fever and obtained similar results.* These protective inoculations have been made upon large numbers of sheep and cattle within the past three years, and with very remarkable success. Recently, however, it has been shown that the protection afforded by the inoculation is a very temporary one, and that after a variable but brief interval, the protected animals are again liable to be fatally attacked by the disease. The opinion seems to be justified that cultivation produces only a temporary degeneration of the bacteria, which rapidly disappears when the organisms are again brought in relation with their proper nutritive fluid. The 'protective inoculations' produce a mild attack of the disease, which is for a time a bar against a second attack, but the effect soon wears off, leaving the animal in its pristine condition of receptivity toward the infective material.

[The following works on this subject are recommended to the student:

STERNBERG AND MAGNIN: The Bacteria; second edition. FLUEGGE: Fermen'e und M kropsrasiten, in VON PETTENKOFER UND ZIEMSSEN's Handbuch d. Hygiene]

* As these sheets are passing through the press, Pasteur reports a like success with the cultivated 'microbe' of rabies.

CHAPTER XVIII.

CONTAGION AND INFECTION.

The adjectives, contagious and infectious, are used to designate certain diseases which are propagated by immediate contact, or through the intervention of some other medium, from the sick to the healthy. The matters in which reside the morbific power, are now believed by many to be vegetable organisms, but not a few pathologists hold to the view that the real contagia, or disease-bearing agents, are modified animal cells or abnormal fluids.

The differentiation between contagion and infection is not easy. Many of the diseases commonly called contagious, are also infectious, that is, they are propagated not merely by direct contact, but also by air, water or food which may have become infected with the morbific agents. Syphilis, for example, may be regarded as simply a contagious disease ; at the present day, at least, we cannot conceive syphilis to be propagated by breathing infected air, or drinking water contaminated with the poison of syphilis. Cholera, typhoid and yellow fevers, on the other hand, are examples of infectious diseases, neither of them being directly contagious, but conveyed from sick to well through the medium of contaminated air, water or food. Between these two stand small-pox and typhus fever, (and perhaps the other exanthemata) which are not merely contagious but infectious also.

There is still a third class of acute diseases, not properly included in either of the classes mentioned. This is the class of miasmatic diseases, of which malarial fevers are the type. According to recent observations, pneumonia ought perhaps to come in this class.

The contagious and infectious diseases are of particular interest to sanitarians, because it is believed that by the judicious carrying-out of sanitary measures they can be prevented. Hence they are sometimes termed preventable diseases. Another peculiarity of the infectious diseases is that they usually occur in groups of cases. Thus small-pox, measles, scarlet fever, typhus fever, diphtheria, and others of the class do not occur sporadically, as it is termed; that is to say, it rarely happens that only one case of small-pox is observed in a locality, unless active measures are at once taken to stamp it out. Usually a number of cases occur successively, and in most instances the succeeding cases can be traced ultimately to the first case.

Contagious and infectious diseases frequently appear as epidemics. Authorities differ as to the proper definition of an epidemic; that is, given the population of a place, how many cases of an infectious or contagious disease are necessary before the disease can be considered epidemic at such place. The following formula was given by the New Orleans Medical and Surgical Association in response to the query: 'Under what circumstances is it proper to declare such diseases (diphtheria, scarlet fever, measles, small-pox, yellow fever, etc.) epidemic in a place?' The answer given is that the disease should be declared epidemic when the number of cases should reach these proportions:*

For a population of	100	- - -	5	per cent.
" " "	500	- - -	4	" "
" " "	2,000 to 5,000 -		22½	" thousand
" " "	6,000 " 10,000 -		16	" "
" " "	20,000 " 50,000 -		8	" ten thousand
" " "	50,000 " 100,000 -		4	" " "
" " "	" 200,000		1	" " "

A disease is said to be *pandemic* when it spreads rapidly over a great extent of country, and *endemic* when it is constantly present in a place. Diseases

* Public Health. Vol. VI, p. 416-417.

which may be prevalent in certain localities, *i. e.*, endemic, not infrequently spread over larger areas of country, overflow their borders, as it were, and become epidemic or pandemic.   Thus cholera, which is endemic in certain districts of India, frequently spreads over adjacent territory, and, at times, the epidemic wave, as it has been called, rolls over nearly the whole world. Plague, malarial and yellow fevers make similar epidemic excursions into other countries, or sections of country, at a distance from the places where they are endemic.

CHAPTER XIX.

HISTORY OF EPIDEMIC DISEASES.

An important part of the knowledge of the sanitarian, is that which relates to the history of the great epidemic diseases that have at various periods devastated large areas of the inhabited world. In this chapter the history of these diseases will be briefly traced. Although some of these diseases have nearly or quite ceased, a knowledge of their habits, and of the causes which finally led to their extinction is of great value, for the reason that the principles and measures of prevention which were effective in times past, are the same which must apply at present and in the future. Hence, time spent in looking back over the fields traversed and noting victories won, will not be wasted.

The epidemic diseases which will here claim attention are, the oriental plague, the sweating sickness, small-pox, Asiatic cholera, typhus, typhoid, scarlet, relapsing, and yellow fevers, diphtheria, dengue, epidemic influenza and syphilis. In addition, some information will be given on certain of the diseases of animals transmissible to man. Among these are sheep pock, actinomycosis, bovine tuberculosis (perlsucht), rabies, anthrax (milzbrand), and glanders.

THE ORIENTAL PLAGUE.

The oriental plague, bubonic plague, the black death, or simply 'the plague,' or great pestilence, overtopping in its fatality all other pestilences, is mentioned by a number of the Greek and Latin medical authors. The first account which clearly refers only to this disease is given by Procopius. According to this

and other contemporary authors, the disease began to
spread in the year 542 from lower Egypt, passing in
one direction along the coast of northern Africa, and
in the other invading Europe by way of Syria and Palestine.   In the course of the succeeding years this pandemic reached 'the limits of the inhabited earth' in
the language of the writers of the day.   The disease
prevailed about half a century, and produced the
greatest devastation wherever it appeared.   'Cities
were devastated, the country converted into a desert,
and the wild beasts found an asylum in the abandoned
haunts of man.'*

The plague is an acute infectious disease, which is
characterised by an affection of the lymphatic system,
i. e., inflammation and swelling of the external and
internal lymphatic glands.   Accessory symptoms are
petechial spots upon the skin, and hemorrhages from
various organs, as the stomach, nose, kidneys, rectum
and uterus.   Those attacked suffer in varied degrees of
intensity.   In some, a fulminant form occurs which
carries off the patient within three days; there is
another class of cases in which buboes develope, with
accompanying fever and hemorrhages, and finally a
light form, rarely fatal, in which only the local symptoms are manifested.   In the great pandemic of the
plague in the fourteenth century, cough and bloody
expectoration were very frequent.   In the later epidemics hemorrhage from the lungs has been rarely
noticed as a symptom.

About the middle of the fourteenth century the
bubonic plague made a second incursion into Europe
from its home in the east.   A most graphic description
of its ravages is given by Boccaccio in the *Decameron*.
This author states that 'between March and July following, according to authentic reckonings, upward of a

* WARNEFRID: quoted by HIRSCH: Hist-Geographische Pathologie, I., p. 350.

hundred thousand souls perished in the city (Florence);
whereas, before that calamity it was not supposed to
contain so many inhabitants.'

This terrible epidemic was forcibly characterised
by its common name, 'the black death.' Hecker estimates that during its continuance, from 1347 to 1351,
25,000,000—one-fourth of the probable total population
of Europe—died. In various cities the mortality was:
in London, 100,000; in Paris, 50,000; in Venice, 100,000;
in Avignon, 60,000; in Marseilles, 16,000 in one month.
It was said that in all England, scarcely a tenth part
of the population escaped death from the disease.

The moral effects of this great pandemic of the
plague were hardly less deplorable than the physical.
Religious fanaticism held full sway throughout Europe,
finding its vent in all manner of excesses. The so-
called Brotherhood of the Cross, otherwise known as
the Order of Flagellants, which had arisen in the
thirteenth century, but had been suppressed by the
ecclesiastical authorities, was revived during the black
pestilence, and large numbers of these religious enthusiasts roamed through the various countries on their
great pilgrimages. Their power increased to such a
degree that Church and State were forced to combine
for their suppression. One consequence of this fanat-
ical frenzy was the persecution of the Jews. These
were accused of being the cause of every evil that befel
mankind, and many were put to death.

In the fifteenth and sixteenth centuries the plague
was generally diffused throughout Europe, and in the
second third of the seventeenth century, its final incursion into the occident took place. The great epidemic
in London, so graphically described by Defoe,* occurred
in 1665. In the early part of the eighteenth century
(1720) the plague visited Marseilles and Toulon; from

* Journal of the Plague in London.

1769 to 1772, it was epidemic in Moldavia, Wallachia, Poland and southern Russia; near the close of the last, and again in the beginning of the present century, in Transylvania, Wallachia, southern Russia and Greece. Very recently, in 1878 and 1879, the plague threatened a new irruption into European territory, being epidemic in the district of Astrachan, on the Caspian sea.*

Although the bubonic plague has never been observed in America, and has spared Europe almost entirely during the present century, it still persists in certain countries of Asia and Africa, especially in Arabia, Mesopotamia, Persia, and the coast of Tripoli.

The older authors ascribed the origin of the plague to various real or supposed conditions. Comets, conjunctions of the planets, 'God's just punishment for our sins,' and similar causes were advanced to account for the outbreaks. Most of the writers of the post-medieval and modern epochs ascribed the disease to meteorological conditions. Observing the fact that the plague never advanced into the torrid zone, and that an epidemic generally ended with the advent of hot weather, a high temperature was believed to be incompatible with the existence of an epidemic, and a cold or temperate climate was considered necessary to an outbreak of the disease. The exceptions to the rule are so numerous, however, that the theory of the climatic or meteorological origin of the plague failed of support. The theory which ascribed the origin of the epidemics to the influence of certain hot and dry winds or a high humidity is also insufficient. Certain geological formations have been supposed to furnish favorable conditions for the development of the disease. Facts show, however, that the disease has prevailed epidemically and endemically in various parts

---

* While these sheets are under press (November, 1884) the journals announce a new outbreak of the plague in Asiatic Russia.

of the earth, and of the most diverse geological character. A certain elevation above sea-level has been held to confer immunity, but recent observations in India show that this belief is unfounded, even places at an elevation of 10,000 feet above sea-level giving no security against attack.

There is, however, one point upon which nearly all writers who mention the fact at all agree. This is that *bad hygienic conditions* are always present where plague prevails. Nearly all observers who have left their impressions on record mention the accumulation of filth in the houses and streets, deficient removal of excrementitious and other sewage matters, crowding and imperfect ventilation of dwellings, as causes favoring the development and spread of the pestilence. All point out the necessity of the removal of these evils as the most important prophylactic measure to be adopted, and all of them further call attention to the fact that those classes of the population most exposed to these unfavorable influences suffered most from the violence of the epidemic.

The later reports of the epidemics in Persia, India, Mesopotamia and Russia, agree in asserting that nothing seems to have promoted the epidemic and endemic prevalence of the plague so much as the material wretchedness of the inhabitants of those countries. In a collection of papers on the plague, printed by a British Parliamentary Commission in 1879, occur these statements: 'The filth is everywhere,' says Mr. Rennie, one of the reporters, 'in their villages, their houses and their persons. Their dwellings are generally low and ill-ventilated,'except through their bad construction ; and the advantage of the natives in other parts of India, of living in the open air, is lost to the villagers of Ghurwal, from the necessity of their crowding together for mutual warmth and shelter against

the inclemency of the weather.' Dr. Dickson, reporting on the plague in Irak Arabi in 1876, says : 'The
most palpable and evident of all the causes which predispose an individual to an attack of plague during an
epidemic outbreak is *poverty*. No other malady shows
the influence of this factor in so striking a degree; so
much so, indeed, that Dr. Cabiadis styles the plague
miseriæ morbis. In his experience (1876-'77, in Bag-
dad) he found that the poor were seldom spared, the
wealthy hardly ever attacked.'*

The manner of transmission of the plague is generally by prolonged inhalation of an infected atmosphere. Hence, it may be termed an infectious disease,
although it is not improbable that it may be communicated by direct contact both of persons and of fomites.

This indicates the measures of prevention to be
adopted. They consist in a rigid quarantine of persons and fomites, prompt and complete isolation of
infected individuals and localities, and destruction (by
fire) or thorough disinfection by steam or sulphurous
acid gas of all materials capable of conveying the virus
of the disease.

THE SWEATING SICKNESS.

This name concisely characterises an epidemic disease, which for the first time appeared in the city of
London and other parts of England in the autumn of
1485. According to Lord Bacon,† the disease began
about the twenty-first of September, and lasted until
near the end of October. It broke out a second time
in the summer of 1507; a third time in July, 1518,
spreading in the course of six months throughout
England. In May, 1529, the disease made its appearance again in the latter country, spreading thence over
a great part of the continent of Europe. Another very

---

* Hirsch, op. cit., p. 370.
† History of Henry VII.

malignant epidemic broke out in the spring of 1551, lasting throughout the summer, and limited in its ravages to England.

With this last outbreak in 1551, this disease disappeared entirely in England and has not reappeared there up to the present day. In the beginning of the eighteenth century, however, a disease very similar in its symptoms and course broke out in Picardy and other districts of northern France, being confined for a number of years to this section of the country. Toward the end of the century, it spread to the south of France, and since that time has appeared epidemically at intervals; 194 distinct outbreaks being observed in the course of 156 years, from 1718 to 1874. The disease has frequently appeared in Italy since 1755, and in various parts of Germany since 1801. In Belgium it has been observed at a few places within the present century.

The disease appeared suddenly; often at night time. The patient was attacked with palpitation of the heart, dyspnœa, great anxiety and oppression, and profuse perspiration. A miliary eruption often appeared on the skin. In favorable cases these symptoms diminished in the course of one or two days, the urinary secretion, which had been suppressed, was restored, and the perspiration became gradually less free. Recovery ensued in from one to two weeks. In the grave cases there were in the beginning of the attack, violent headache, delirium, convulsions, followed by a comatose condition, from which the patients rarely recovered.

This disease is undoubtedly of a miasmatic infectious nature, as proved by its rapid spread, and limitation to certain localities. It appears most frequently in the spring and summer, and is nearly always observed in marshy or damp localities. Its spread is

favored by a high temperature and humidity. There is
no apparent connexion between the outbreaks of the
sweating sickness and overcrowding or other insanitary
conditions; in fact, it is stated by numerous observers,
both old and recent, that children, the aged, and
generally the poorer classes were remarkably exempt
from the disease.

Since the first appearance of Asiatic cholera in
France, in 1832, an apparently intimate connexion has
been observed between the occurrence of that disease
and outbreaks of sweating sickness. A disease strongly
resembling the sweating sickness has also been observed
in India in districts contiguous to places where cholera
was at the time epidemic.[*]

SMALL–POX.

The earliest details concerning small-pox are derived from certain Chinese records, according to which
it appears that this disease was known in China upward
of two thousand years ago. It was also known at a
very early period in India. It is believed to have been
introduced into Europe in the second century, by a
Roman army returning from Asia. It is believed that
the emperor Marcus Aurelius died of small-pox, which
prevailed in his army at the time of his death.

The first distinct references to small-pox in medical
literature occur in the writings of Galen, in the second
century. Rhazes, in the ninth century, wrote upon the
disease, describing it very accurately.

The almost universal susceptibility to small-pox
caused wide-spread devastation wherever it appeared,
previous to the introduction of vaccination. The statement is made that in England, in the last century,
about one person in every three, was badly pock-
marked. The mortality from the disease was exceed-

---

[*] MURRAY; Madras Quart. Med. Journ. 1840-41. Quoted in HIRSCH; l. c. p. 83.

ingly great, being, in the latter half of the eighteenth century, about 3,000 per million of inhabitants annually.

In India the mortality from small-pox has been exceedingly great within the last twenty years. From 1866 to 1869, 140,000 persons died in the Presidencies of Bombay and Calcutta, having a population of about forty millions. Several years later, from 1873 to 1876, 700,000 died from this disease.

China, Japan, Cochin China, the islands of the China sea, and Corea are frequently ravaged by small-pox. In the latter country nearly all the inhabitants are said to bear evidence of an attack of the disease.

The Samoyedes, Ostiaks and other natives of eastern Siberia have frequently suffered from devastating epidemics. In Kamtchatka, the disease was introduced in 1767, and produced frightful ravages. Many villages were completely depopulated.

In Mexico, small-pox was introduced by the Spaniards in 1520. In a short time, it carried off over three and a half millions of the natives. In the Marquesas islands, one-fourth of the inhabitants have fallen victims to the disease since 1863.

It was first introduced into the Sandwich islands in 1853, and carried off eight per cent. of the natives.

Australia, Tasmania, New Zealand and the Fejee archipelago remain exempt to the present day from small-pox. It has several times been carried to Australia by vessels, but has always been promptly checked by the vigilance of the authorities.

On the western hemisphere small-pox was unknown before the arrival of the European conquerors. It has been spread by the whites, or imported African slaves, to nearly all the Indian tribes of both continents. When it attacks large communities unprotected by previous out-breaks of the disease, or by inoculation, or vaccination its ravages are frightful. The mortality of unmodified small-pox is usually between thirty and forty per cent.

Small-pox is a highly contagious and infectious disease. It is produced by actual contact, by inoculation, and by inhaling an atmosphere charged with the poison. In order to cause an outbreak two factors are necessary: first, a number of individuals susceptible to the disease; and second, the introduction into the body, in some manner, of the virus upon which it depends.

Small-pox is spread from: 1, persons sick with the disease; 2, others, not themselves sick, or susceptible, but coming in contact with the poison; 3, fomites, (cotton, wool, etc.) and 4, the bodies of persons dead with small-pox. It is also probable that the air in the immediate vicinity of a person sick with small-pox becomes charged with the poison and able to convey the disease. It is at present impossible to fix the distance to which this infectiousness of the air extends, but it does not ordinarily reach beyond the room in which the patient is confined.

It is a fact of common observation that the darker races are more commonly attacked, and the disease is likewise more fatal among them. The mortality among negroes is much larger than among other races.

It is a current belief that small-pox is only contagious after the development of the pustules. This is a serious error. It is probably contagious in all stages of the disease ; certainly as early as the first appearance of the eruption, and probably even in the stage of preliminary fever.

One attack of small-pox usually confers immunity from the disease for life. This rule has its exceptions however, which, if not numerous, are yet not infrequent. The author has seen a case in which the patient suffered from a third attack of the disease.

The popular belief, that persons suffering from any other acute or chronic disease are less liable to be attacked by small-pox than those at the time in good health, is erroneous. On the contrary, the subjects of chronic disease, such as consumption or dyspepsia, are much more liable to succumb to an attack of small-pox, than persons previously in good health.

It is true, however, that individuals suffering from some other acute infectious disease, like scarlet fever, measles, typhoid fever, etc., are generally, though not absolutely, exempt from an attack of small-pox during the time they are sick with such other disease. But if they are exposed, after recovery, to the small-pox infection, their liability to an attack is as great as in those who have not passed through a similar disease. A number of cases have been reported by Curschmann,* in which infection by small-pox took place on the day in which convalescence from typhoid fever was established.

The author has reported a case,* in which the patient passed through an attack of erysipelas, during the incubative stage of small-pox. From all the evidence attainable, the incubative stage was not prolonged by the intercurrent erysipelas.

Epidemics of small-pox usually begin in the autumn or winter, and lessen in violence as warmer weather approaches. The spread of the disease is slow at first, increasing in rapidity as the foci of infection multiply.

When the poison is imported into a community late in the spring, or during the summer, the increase in the number of cases is exceedingly gradual until colder weather sets in. If it is introduced during the winter, the disease spreads much more rapidly, but

* ZIEMSSEN's Cyclopædia, Vol. II.
* Medical News, July 7, 1883.

decreases, and sometimes almost disappears during the summer. On the return of cold weather, however, the epidemic starts out with a new lease of activity and presents great difficulties to its restriction.

Inoculation.—The prevention or restriction of such a universal and fatal pestilence as small-pox, is a matter of the deepest importance. The first attempt to limit its fatality dates from the end of the seventeenth century. It became generally known in Europe about the year 1700, that the intentional inoculation of variolous matter into healthy individuals induced an attack of the disease which generally ran through its various stages with less virulence than when the disease was contracted in the usual manner. In 1716 and 1717, two papers were published in the Transactions of the Royal Society of England giving an account of the process of inoculation. The attention of the public was especially directed to the matter, however, by the famous letter of Lady Mary Wortley Montagu, dated April 1, 1717. This letter is as follows:* 'Apropos of distempers, I am going to tell you a thing that will make you wish yourself here. The small-pox, so fatal and so general amongst us, is here entirely harmless, by the invention of *ingrafting*, which is the term they give it. There is a set of old women who make it their business to perform the operation every autumn, in the month of September, when the great heat is abated. People send to one another to know if any of their family has a mind to have the small-pox; they make parties for this purpose, and when they are met—commonly fifteen or sixteen together—the old woman comes with a nut-shell full of the matter of the best sort of small-pox, and asks what veins you please to have opened. She immediately rips open that you offer to her with a large needle—which gives you no more pain than a

---

* The letter is addressed to Mrs. S. C. (Sarah Chiswell).

common scratch—and puts into the vein as much matter as can lie upon the head of her needle, and after that binds up the little wound with a hollow bit of shell; and in this manner opens four or five veins. The Grecians have commonly the superstition of opening one in the middle of the forehead, one in each arm, and one on the breast, to make the sign of the cross; but this has a very ill effect, all these wounds leaving little scars, and is not done by those that are not superstitious, who choose to have them in the leg, or that part of the arm that is concealed. The children or young patients play together all the rest of the day, and are in perfect health until the eighth. Then the fever begins to seize them, and they keep their beds two days, very seldom three. They have very rarely above twenty or thirty in their faces, which never mark; and in eight days' time they are as well as before their illness. Where they are wounded, there remain running sores during the distemper, which I don't doubt is a great relief to it. Every year thousands undergo this operation; and the French ambassador says pleasantly, they take the small-pox here by way of diversion, as they take the waters in other countries. There is no example of any one that has died in it, and you may believe that I am well satisfied of the safety of the experiment since I intend to try it on my dear little son.

'I am patriot enough to take pains to bring this useful invention into fashion in England; and I should not fail to write to some of our doctors very particularly about it, if I knew any one of them that I thought had virtue enough to destroy such a considerable branch of their revenue for the good of mankind. But that distemper is too beneficial to them, not to expose to all their resentment the hardy wight that should undertake to put an end to it. Perhaps, if I live to return, I may however have courage to war with them.'

Soon after the date of this letter, the writer's son was inoculated in Turkey, and four years later her daughter also, being the first subject inoculated in England. The practice soon became popular, but several fatal cases among prominent families brought it into disrepute, and for about twenty years very few inoculations were made in England. It was revived about the middle of the century by the founding of a small-pox and inoculation hospital in London. This continued in operation until 1822. The records of this institution showed that only three in a thousand died of the disease thus communicated. The practice has now fallen into desuetude, being superseded by vaccination, and prohibited by law in England.

Inoculation was introduced into this country in 1721 by Dr. Zabdiel Boylston, of Boston, who had his attention directed to the practice by Cotton Mather, the eminent divine.* During 1721 and 1722, 286 persons were inoculated by Boylston and others in Massachusetts, and six died. These fatal results rendered the practice unpopular, and at one time the inoculating hospital in Boston was closed by order of the Legislature. Toward the end of the century an inoculating hospital was again opened in that city.

Early in the eighteenth century inoculation was extensively practiced by Dr. Adam Thomson, of Maryland, who was instrumental in spreading a knowledge of the practice throughout the Middle States.†

In China and India, and perhaps other eastern countries, inoculation was practiced at a very early period.

The inoculation of variolous matter, although it mitigated to a very great degree the attack of small-

---

* Dr. Jno. R. Quinan, (*Md. Med. Journ.*, June 23 and 30, 1883.) believes the claim of Dr. Boylston to be the first American inoculator open to question. The evidence presented is, however, insufficient to discredit the claim of the Boston physician.

† See Quinan, l. c., p. 114.

pox following, had one very serious objection, aside from the small death-rate which was a direct consequence of it. This was the fact that inoculation always produced small-pox, and thus assisted in propagating the disease, for however mild the induced disease might be, the inoculated individual was liable to communicate small-pox to others in its most virulent form. Hence, nothing short of universal inoculation, which was manifestly impracticable, would succeed in reducing the danger from the disease.

Vaccination.—It had been noticed at various times that a pustular disease which sometimes appears on the udders of cows, called cow-pox, had not infrequently been transmitted to the hands of dairy-maids and others having much to do with cows. In course of time it was also noticed that persons who had been thus attacked never suffered from small-pox. This protective power of cow-pox was known as early as 1713, and in 1774 Benjamin Jesty, a Gloucestershire farmer, performed vaccination for the first time on record, inoculating his wife and two sons with cow-pox matter as a protection against small-pox.

It is stated that when it became known that Jesty had vaccinated his wife and sons, 'his friends and neighbors who had hitherto looked upon him with respect, on account of his superior intelligence and honorable character, began to regard him as an inhuman brute, who could dare to practice experiments upon his family, the sequel of which would be, as they thought, their metamorphosis into horned beasts. Consequently the worthy farmer was hooted at, reviled and pelted whenever he attended the markets in his neighborhood.'*

In 1791 a school teacher in Holstein also inoculated three boys with the matter of cow-pox, but nothing is known of the subsequent history of these cases.

---

* London Lancet, September 13, 1862.

Although the above facts are clearly established, it is to Edward Jenner, a modest country doctor of Berkeley, in the county of Gloucester, in England, that the merit of demonstrating the protective power of cow-pox against small-pox, and of diffusing a knowledge of this fact, is due. Jenner had his attention directed to the asserted protection conferred by cow-pox during the period of his apprenticeship. After a residence in London as a pupil of John Hunter he returned to the country to practice his profession. About the year 1776 he began studying the question, and gathering evidence of the protection afforded against small-pox by the accidental inoculation of cow-pox virus. For twenty years he studied the subject, patiently awaiting an opportunity to put his belief to the test of experiment. On the fourteenth of May, 1796, he made his first vaccination on a boy named James Phipps. Six weeks later he inoculated this boy with variolous matter, but without success, no small-pox resulting. Two years later he published his pamphlet, entitled 'An Inquiry into the Causes and Effects of the Variola Vaccinæ, etc.,' in which he detailed his observations and experiments. This publication produced a great sensation in the medical world, and although much opposition was at first manifested toward his views, he soon gained many adherents.

Vaccination, as the operation for the inoculation of cow-pox virus is termed, was rapidly introduced into all civilised countries, and soon demonstrated its good effects by greatly restricting the prevalence of small-pox. It is generally believed that the first one to practice vaccination in this country was Dr. Benjamin Waterhouse, of Boston, in the summer of 1800, but Dr. John R. Quinan has recently shown* that vacci-

* QUINAN, l. c., p. 118, 131.

nation was introduced into Maryland by Dr. John
Crawford and Dr. James Smith, at least as early as the
date generally assigned for its introduction into
Massachusetts.

It was believed by Dr. Jenner, and was afterward
conclusively shown by a number of distinguished
experimenters, that vaccinia, as the disease produced
by cow-pox inoculation was called, was merely a modification of small-pox, as it existed in the cow. Small-
pox virus when inoculated upon the cow, produced
cow-pox; but the latter reinoculated upon man,
produced cow-pox (vaccinia), and not small-pox.
Sheep-pock and horse-pock or 'grease,' are probably
merely modifications of the disease produced by inoculating small-pox into those animals.

When cow-pox virus is successfully inoculated into
the human system, that is, when a person is successfully vaccinated, the following local and general
symptoms are observed:

In the case of a primary vaccination, i. e., where
the individual has not been previously vaccinated, or
attacked by small-pox, the point where the vaccination
is made, shows no particular change for the first two
days. If the vaccination is successful, a small, reddish
papule appears by the third day, which, by the fifth or
sixth day has become a distinct vesicle of a bluish
white color, with a raised edge, and a peculiar, central,
cup-like depression, called the umbilication. By the
eighth day this vesicle has become plump, round, and
pearl-colored, the central umbilication being still more
marked. At this time a red, inflamed circle, called the
areola, appears, surrounding the vesicle and extending
usually in a radius of from one-half to two inches
when fully developed. This inflammatory ring is
usually pretty firm, and there is more or less general
fever, and often enlargement and tenderness of the

axillary glands.   After the tenth day, the areola begins
to fade, and the contents of the vesicle dry into a hard,
brownish crust or scab, which falls off between the
twentieth and twenty-fourth day, leaving a punctated
scar, which gradually becomes white.

When the vaccinia has passed through all of these
stages, especially if the vesicle filled with pearly lymph,
and the areola have been well developed, the vaccination may be considered a success, and the individual
protected against small-pox for a number of years, if
not for life.   Recently the doctrine has been strongly
advocated that vaccination is not absolutely protective
until a subsequent inoculation of vaccine fails to 'take.'
According to this view, vaccination should be repeated
until it fails any longer to exhibit any local reaction.
When this has been attained the individual may be
considered absolutely protected for life.   Theoretically,
this view has much in its favor, but there is as yet not
sufficient evidence to establish it as a law.

It may be stated as an established fact that vaccination, although carefully performed and successful,
does not confer absolute immunity from small-pox for
life.   The protective power seems to wear out after a
time and the individual then again becomes susceptible to small-pox.   An attack of small-pox in a
vaccinated individual is, however, nearly always much
milder than where there had been no vaccination.
There is no fact in the entire range of medicine better
established than this: that small-pox in vaccinated
persons is a much less dangerous disease than typhoid
fever, while in unvaccinated cases the mortality ranges
from thirty to forty per cent.   An approximate guide
to the beneficent influence of vaccination upon the
mortality from small-pox is furnished by a table in
Seaton's report on vaccination.   Before the introduction of vaccination, the mortality from small-pox per

million of inhabitants of England was nearly 3,000 per year. After the introduction of vaccination, the mortality was reduced to 310 per million per year.

The most remarkable and convincing statistical evidence on the question is given by Drs. Seaton and Buchanan of England. During the small-pox epidemic in London in 1863, they examined over 50,000 school children, and found among every thousand without evidence of vaccination, 360 with scars of small-pox; while of every thousand presenting some evidence of vaccination, only 1.78 had any such traces of small-pox to exhibit.* The reliability of general mortality statistics may be called in question—in some cases with justice; but the significance of these figures cannot be evaded.

The upper and outer surface of the arm is usually chosen as the point where the virus is inserted, although any part of the body, which can be protected against friction, or other mechanical irritation, may be selected. The method varies slightly in the hands of different vaccinators. The two methods most frequently in use are scarification and erasion. The former method has the endorsement of Mr. Seaton, the high English authority. The method of erasion—scraping off the epidermis until the papillary layer of the skin is laid bare—is now most frequently used in this country. The best instrument to use is a clean thumb lancet; in default of this, an ordinary sewing needle answers well. Where animal vaccine is used, the ivory slip, or sharpened quill may also be used with satisfaction to make the scarification or erasion. Whatever instrument is used, it should always be kept perfectly clean.

A point of vital importance is that which relates to the proper age at which children should be vaccina-

* SEATON: Vaccination, in REYNOLDS' System of Medicine, Second Edition, Vol. I. p. 291.

ted. Ordinarily vaccination should be performed within the first six months of life. In time of danger from a threatened, or in the presence of an actual epidemic, infants may be vaccinated when only one day old.

In order to secure permanent protection against small-pox, revaccination should be performed after a certain interval. Some place the period at which this second vaccination should be done at five years, while others allow a longer interval—seven, eight, or ten years. The law of Prussia is that every child that has not already had small-pox must be vaccinated within the first year of its life, and every pupil in a public or private institution is to be revaccinated during the year in which his twelfth birthday occurs.

A revaccination, even if successful, rarely passes through all the typical stages of a primary vaccination. The vesicle rarely becomes so full and plump, and is more frequently flat and irregular in outline. Swelling of the axillary glands and other complications also seem to be more frequent than in cases where the vaccination is done for the first time.

The question, whether the lymph direct from the cow, or humanised lymph is the more efficient, has caused much discussion. The objections urged against the use of humanised virus are : first, that its protective power has become diminished by transmission through many generations ; second, that it is liable to transmit other diseases, such as syphilis, tuberculosis, scrofula, etc., third, that it is frequently difficult to obtain in sufficient quantities in an emergency, such as an actual or threatened epidemic.

The first objection is disproved by the testimony of many of the most distinguished medical men in Europe and this country. Humanised vaccine virus, when properly inoculated, seems to be as completely protective against small-pox, as that taken direct from the animal. Among its advantages are that it 'takes' more readily, and runs through its stages of development in a shorter time, and that it will retain its active properties for a greater length of time than animal virus. The physician can usually control the source whence he obtains it. He can watch over the subject that furnishes it and reject that which is suspicious. With humanised lymph collected by the physician himself, there need be no doubt as to its purity or age ; with animal lymph furnished by the cultivators of that article, there can be no certainty about either of these important points.

That syphilis has been inoculated with humanised vaccine virus can no longer be open to doubt. The recent experiment of Dr. Cory, of England, has settled this question definitely. With care, however, this sad accident can easily be avoided, and the fact that syphilis has been so rarely transmitted by vaccination is sufficient evidence that the danger of such infection is not very great.

The most serious objection against the exclusive use of humanised lymph, is, that in grave emergencies, such as a rapidly spreading epidemic of small-pox, it is difficult to obtain a sufficient supply of the lymph.

Humanised virus is inoculated either in the fresh state, i. e., the lymph is taken from the vesicle on the seventh day, and inoculated directly into the arms of other individuals ; or else, the vesicle is allowed to dry into a crust, with which a thin paste is made by moistening with water at the time of vaccination. The readiest way of using the crust is, to crush a small fragment between two small squares of glass, then moistening it with a drop of warm (not hot) water, and smearing it on the spot where the vaccination is to be made. With the lancet a number of cross scarifications are then made, and the virus well rubbed in. Only so much of the crust should be moistened as will be used at the time. Particular care must be taken not to use saliva for moistening the crust. Aside from being unclean, there is danger of producing blood-poisoning by inoculating certain of the oral secretions.

Animal virus is obtained by vaccinating a calf or heifer with vaccine virus, either derived from a case of small-pox, from another case of cow-pox or by reinoculating humanised virus into the animal. The vesicles are opened on the seventh day, and ivory points or the ends of quills coated with the lymph, and dried with a gentle heat.

In vaccinating with animal virus, the quill or ivory point is first moistened with a drop of water to soften the adhering lymph, the scarification or abrasion of the skin is then made with the lancet, and the virus rubbed well into the scarified spot.

In using animal virus the successive stages of development are usually one or two days later than when humanised virus is used. In the former case the areola is rarely developed before the ninth day.

Certain complications are likely to occur in the course of the vaccinia of which the student should be aware.

When the areola appears there is usually more or less fever. Sometimes the constitutional manifestations are decidedly marked, fever of a high grade being not uncommon. In addition to the glandular enlargement and tenderness, an outbreak of roseola sometimes comes on about the ninth or tenth day. This eruption may be mistaken for scarlet fever, but if it is remembered that two infectious diseases rarely co-exist in one individual during their full development, this error will be avoided.

Erysipelas involving the entire arm is sometimes observed as a complication of vaccination. This, in nearly every case, depends upon some depravement of the patient's constitution, innutrition, bad sanitary surroundings, or perhaps, more frequently, chronic alcoholism. Individuals who are habitually intemperate in the indulgence of alcoholic liquors, are especially unfavorable subjects for vaccination. The results are, fortunately, rarely serious to the patient.

Another inconvenient complication of vaccination is the formation of a deep, ill-looking, sloughing ulcer at the vaccinated point. This has been, in the author's experience, a much more frequent concomitant when animal virus has been used, than when humanised virus was resorted to. It should be borne in mind that a very sore arm, especially if followed by the formation of an ulcer, or gangrenous sore, may not be protective against small-pox. Such a patient should not be considered properly vaccinated, and must be revaccinated as soon he recovers, or immediately, if there is danger of small-pox infection.

Children with eczematous eruptions, however localised upon any portion of the body, should not be vaccinated until the eruption is first cured, except in times of danger from small-pox. The eczema will be almost certainly rendered worse in consequence of the general hyperemia accompanying the febrile reaction, and the physician who performs the vaccination will be blamed for causing the skin disease.

The author has placed on record* two cases of general psoriasis following vaccination, and other cases have been since reported. Urticaria and exudative erythema has also been repeatedly observed.

As before stated, syphilis may be communicated to the vaccinee by vaccine virus obtained from a syphilitic subject, but this accident is infrequent. There can be little doubt that some of the cases reported

*Journal Cutaneous and Ven. Diseases. Vol. I, No. 1, p. 11.

as 'vaccinal syphilis,' are cases of tardy hereditary syphilis, lighted up by the general systemic disturbance following vaccination.

Next in importance to vaccination in the prophylaxis of small-pox, is prompt isolation of the sick. No one but the medical and other attendants of the sick should be allowed to come in contact with them. All attendants and other persons exposed to the infection should, of course, be promptly vaccinated, unless this has been successfully done within the previous year or two.

Disinfection of all discharges from the patient and of the room and its contents after the patient has recovered or died, must be practised. The best disinfectants in small-pox are bichloride of mercury, free chlorine, and sulphurous acid.

When it is learned that a person has small-pox, if he is not removed to a special hospital, a room should be prepared for his occupancy. The carpets should be taken up and the floor kept clean. Window curtains and unnecessary furniture and drapery should be removed from the room. After recovery of the patient the bed-clothing must be thoroughly disinfected with steam or sulphurous acid, or destroyed by fire.

ASIATIC CHOLERA.

The history of a disease which causes the death of three-fourths of a million of human beings in the country where it is endemic, within the space of five years, and which makes periodical excursions, spreading over the entire globe with destructive violence, must surely command the interest of every intelligent person. The fact that one of these periodical advances of this disease from its home in the far east is even now in progress may lend additional interest to the following pages.

Asiatic cholera is an endemic disease of India, where it probably originated centuries ago. Some authors claim to have found satisfactory evidence of its existence in the writings of the classical authors of India and Greece, at a period as early as the second century of the Christian era. The evidence is however not beyond question. The earliest reliable account of cholera as it exists in its home is contained in the work of a French author.* He describes a pestilence having all the characters now recognised as belonging to Asiatic cholera, which prevailed in the neighborhood of Pondicherry and the Coromandel coast in 1768 and 1769, and which carried off 60,000 of those attacked by it within a year. Dr. Macpherson gives numerous references, which indisputably establish the endemic existence of cholera in India anterior to the beginning of the present century.

Being endemically prevalent over a greater or less area in India for many years, cholera finally, in 1817, crossed the boundaries of that country, and advancing in a south-easterly direction, invaded Ceylon, and the Sunda islands in 1818. In a westerly direction the disease was carried to the islands of Mauritius and Réunion, and finally reached the east African coast in 1820. During this year it also traveled north-easterly devastating the Chinese empire for the two following years, reaching Nagasaki in Japan in 1822.

In 1821, the disease spread from India in a westerly direction, extending along the east coast of Arabia, to the border of Mesopotamia and Persia. In the spring of 1822, it began with renewed violence, following the river Tigris to Kurdistan, and extending further in a westerly direction, reached the Mediterranean coast of Syria. In the following year, 1823, it extended from Persia into Asiatic Russia, reaching

---

* SONNERAT: Voyage aux Indes Orientales, Paris, 1782.

Astrachan on the European border in September, but dying out nearly everywhere beyond the borders of India during the ensuing winter.

In 1826, cholera again advanced from India, reaching Orenburg in Russia, in 1829, and in the following year appeared in St. Petersburg. Extending to the north and south it invaded Finland and Poland in the same year. From Persia, the disease spread to Egypt and Palestine in 1830-31.

From Russia, the pestilence invaded Germany in 1831, passing thence in 1832 into France, the British isles, Belgium, the Netherlands, Norway and Sweden. In the latter year, cholera crossed the Altantic ocean for the first time, being carried to Canada by emigrants from Ireland; and spreading thence to the United States by way of Detroit. In the same year it was imported into New York by emigrants, and rapidly spread along the Atlantic coast. During the winter of 1832, it appeared at New Orleans, and passed thence up the Mississippi valley. Extending into the Indian country, causing sad havoc among the aborigines, it advanced westward until its further progress was stayed by the shores of the Pacific Ocean. In 1834, it reappeared on the east coast of the United States, but did not gain much headway, and in the following year New Orleans was again invaded by way of Cuba. It was imported into Mexico in 1833. In 1835 it appeared for the first time in South America, being restricted however to a mild epidemic on the Guiana coast.

While the pestilence was advancing in the western hemisphere, it also spread throughout southern Europe, invading in turn Portugal, Spain and Italy.

Extending in an easterly direction from India, the disease reached China and Japan in 1830-31; westwardly, Africa was invaded in 1834, and ravaged by the epidemic during the following three years.

This second extensive outbreak of cholera ended in 1837, disappearing at all points beyond the borders of India. In 1846, it again advanced beyond its natural confines, reaching Europe, by way of Turkey, in 1848. In the autumn of this year, it also appeared in Great Britain, Belgium, the Netherlands, Sweden, and the United States, entering by way of New York and New Orleans. In the succeeding two years, the entire extent of country east of the Rocky mountains was invaded. During 1851 and 1852, the disease was frequently imported by emigrants, who were annually arriving in great numbers from the various infected countries of Europe. In 1853 and 1854, the disease again prevailed extensively in this country, being however, traceable to renewed importation of infected material from abroad. In the following two years it also broke out in numerous South American States, where it prevailed at intervals until 1863.

Hardly had this third great pandemic come to an end, before the disease again advanced from the Ganges, spreading throughout India, and extending to China, Japan and the East India archipelago during the years 1863 to 1865. In the latter year it reached Europe by way of Malta and Marseilles. It rapidly spread over the continent, and in 1866 was imported into this country by way of Halifax, New York and New Orleans. This epidemic prevailed extensively in the western States, but produced only slight ravages on the Atlantic coast, being kept in check by sanitary measures. In the same year 1866, the disease was also carried to South America, and invaded, for the first time, the States bordering on the Rio de la Plata, and the Pacific coast of the continent.

While the epidemic was thus advancing westward from its home in India, it was at the same time spreading northwardly over the entire western part of Asia, and in a south-westerly direction over northern Africa. In the latter continent it prevailed from 1865 to 1869.

Cholera never entirely disappeared in Russia during the latter half of the sixth decade, and in 1870 it again broke out with violence, carrying off a quarter of a million of the inhabitants before dying out in 1873. It spread from Russia into Germany and France, and was imported in 1873 into this country, entering by way of New Orleans, and extending up the Mississippi valley. None of the Atlantic coast cities suffered from the epidemic in 1873, and since that year, the United States have been entirely free from the disease.

In 1883, a new epidemic of cholera broke out in Egypt, where it raged with great violence. In 1884 it was imported into France, extending thence into Italy and Spain. The origin and further progress of this epidemic, which is still raging at time of writing, (November, 1884) cannot yet be written.

This brief historical sketch of all the epidemics of cholera observed beyond the borders of India demonstrates several facts: first, that the home or breeding place of cholera is in India, whence it spreads at intervals throughout the world; second, that it always advances along the lines of travel of large bodies of human beings; third, that it advances, by preference, along water routes. The latter is particularly noticeable in the behavior of cholera epidemics in this country. When it has been spread from an Atlantic port, it has generally been to other places having water communication with that port. It seems to spread with difficulty along lines of railroad. When the disease has spread from New Orleans, it has always been up the Mississippi valley, expending its violence on the river cities: Vicksburg, Memphis, St. Louis and Cincinnati.

Several factors must concur before there can be an epidemic of cholera. These are, first, the cholera poison; second, certain local conditions of air, soil, or water; third, individual predisposition. Without a concurrence of all these conditions, no outbreak can occur. If by any means the co-existence of these three conditions can be prevented, cholera can be prevented. The following are facts bearing upon this question:

Cholera is communicated through the agency of a specific poison. This does not admit of doubt. The poison may be either organic germ, inorganic particulate, or gaseous. The recent researches of Dr. Robert Koch indicate that a micro-organism found in the intestinal canal of cholera patients, and termed by him the 'comma bacillus,' (really a spirillum, not a bacillus) is the active agent in propagating the disease. While cholera is not personally contagious in the same sense that small-pox is contagious, there can be no doubt that it is only spread by the poison from other cases of the disease. The regularity of its march along routes by which the intercourse of human beings takes place, and always in connexion with other cases of cholera proves this. There is no undoubted case on record where genuine cholera has been spontaneously developed outside of India.

That certain, local geological and perhaps meteorological conditions are necessary for the propagation or virulence of the poison of cholera is beyond dispute. In nearly all epidemics, certain cities or towns, or portions of a town, into which persons sick with cholera are brought, and where the poison of the disease is thus imported, remain exempt from the effects of the epidemic. The inference to be drawn from this fact is that in such localities the local conditions are unfavorable to the development of the poison and it becomes inert.

In India all the local conditions favorable to the propagation of the cholera are found. The filthy personal habits of the people, the over-crowding, the intense heat, the lack of sufficient or properly prepared food, and the extensive pollution of the water supply, all combine to produce the necessary conditions of development of the poison. These conditions doubtless, to a considerable extent, give rise to that depression of the system which seems necessary to constitute the individual predisposition to become infected.

Given, then, at any place, a number of persons of a lowered degree of vitality ; *i. e.*, not capable of resisting unfavorable influences under unfavoring conditions ; given conditions of surroundings more or less similar to those present in India, only the introduction of the third factor, the cholera poison, is needed to cause an outbreak. In many cities of this country, and of Europe, as proven by the most recent epidemic in Toulon, Marseilles and Naples, the conditions are present which would furnish the most favorable nidus for the cholera germ if introduced.

The dejections and vomited matters of cholera patients, contain the active agent which produces the disease. These excreta are however not infective at the time when passed, but must first undergo certain changes, before they can reproduce the disease.

The cholera epidemic in this country in 1873, furnished a good opportunity for a thorough investigation, which was made by order of Congress, under the supervision of the late Surgeon-General Woodworth, of the Marine Hospital Service, assisted by Dr. Eli McClellan, Surgeon U. S. Army.* The conclusions were put in the form of the following propositions, which may be regarded as embracing the facts within our knowledge at the present time :

---

* Cholera Epidemic of 1873. Washington: Government Printing Office.

I. That Asiatic cholera is an infectious disease resulting from an organic poison, which, gaining entrance into the alimentary canal, acts primarily upon and destroys the intestinal epithelium.

II. That the active agents in the distribution of the cholera poison are the dejections of persons suffering from the disease in any of its stages. That in these dejections there exists an organic matter, which, at a certain stage of decomposition, is capable of reproducing the disease in the human organism to which it has gained access.

III. That cholera dejections coming in contact with, and drying upon any objects, such as articles of clothing, bedding, and furniture, will retain indefinitely their power of infection. That in this manner a sure transmissibility of the cholera infection is effected, and that a distinct outbreak of the disease may occur by such means at great distances from the seat of original infection.

IV. That the specific poison which produces the disease known as cholera, originates alone in India, and that by virtue of its transmissibility through the persons of infected individuals, or in the meshes of infected fabrics, the disease is carried into all quarters of the world. That cholera has never yet appeared in the western hemisphere until after its route of pestilential march which had commenced in the eastern world, and that its epidemic appearance upon the North American continent has invariably been preceded by the arrival of vessels infected with cholera. sick, or laden with emigrants and their property from infected districts.

V. That the respiratory and digestive organs are the avenues through which individual infection is accomplished; that through the atmosphere of infected localities, cholera is frequently communicated to individuals; that water may become contaminated with the specific poison of cholera from the atmosphere, from surface washings, from neglected sewers, cesspools, or privies, and that the use of water so infected will produce an outbreak of the disease.

VI. That the virulence of a cholera demonstration, the contagion having been introduced into a community, is influenced by the hygienic condition of the population, and not by any geological formation upon which they may reside.*

VII. That one attack of cholera imparts to the individual no immunity to the disease in the future, but that the contrary seems to be established.

The prophylaxis against cholera comprises such measures as prevent the admission of the cholera poison into a community; prevention of the development of the poison after its introduction, and the reduction of individual susceptibility to attack.

It is evident from the foregoing, that if the introduction of the cholera poison could be prevented, no outbreak of the disease could occur. With this in view, some have urged the enforcement of a strict policy of non-intercourse with infected localities. At the present day, few sanitarians advocate these extreme measures. A modified system of restricted intercourse is supported by many authorities, who claim that by the adoption of a thorough system of maritime inspection, disinfection, and observation, the poison can be rendered ineffective, or its entrance into a community prevented.

The best authorities, however, think that it is not only easier, but far more effective, to place the threatened locality in such a sanitary condition that the de-

---

*This proposition by no means controverts the view of PETTENKOFER and his followers, that the movements of the ground-air and ground-water are intimately connected with the propagation of cholera, as of typhoid fever. The ground-water theory has been referred to in Chapter IV., p. 96, et. seq.

velopment of the cholera poison cannot take place.
The contrast between the effectiveness of quarantine
and local sanitation as preventives of cholera, has
been well expressed by Pettenkofer, who compares
cholera epidemics to powder explosions.   The virus of
cholera, he says, is the spark that evades the strictest
quarantine.   The powder is the *ensemble* of local conditions which predispose to the outbreak.   'It is,
therefore, wiser to seek out and remove the powder,
than to run after and try to extinguish each individual
spark before it drops on a mass of powder, and ignit-
ing it, causes an explosion which blows us into the air
with our extinguishers in our hands.'

The measures of local sanitation are such as will
secure cleanliness of person, of habitations and surroundings, of air, of water, and of soil.   Pollution of
the soil should be especially guarded against, for a
polluted soil means impure air and water, and these
mean, if not an infectious disease, at least a heightened
receptivity to its influence.

The individual predisposition to cholera is best
guarded against by keeping the body clean and well-
nourished.   Under-feeding, anxiety, over-work, exposure to extremes of temperature, intemperance in eating and drinking should all be avoided, as they tend to
reduce the resistance of the system to the influence of
the morbid poison.

Certain measures of personal prophylaxis which
have proven of use heretofore, should be adopted wherever cholera prevails.   One of these is the use of sulphuric acid lemonade.   Ten to fifteen drops of dilute
sulphuric acid in a glass of water, sweetened with sugar,
may be drunk instead of water.   Experience with it
during the epidemic of 1866, has demonstrated its great
value as a preventive of cholera.

A painless diarrhœa, called cholerine, attacks many persons during cholera epidemics. This disorder is easily curable if promptly attended to, but if allowed to run on, it may develope into a malignant attack of cholera.

Among the means of securing prompt treatment of the poorer classes in times of epidemic is the establishment of numerous public dispensaries, where medical aid can be always obtained. The establishment of such dispensaries, and, if possible, of temporary hospitals in the crowded portions of cities, is a very important part of the prophylactic treatment of cholera.

Inasmuch as it seems definitely established that the discharges from the stomach and intestines are the active agents in propagating the disease, the immediate disinfection of such discharges is vitally important. The stools and vomited matters must be rendered innocuous by germicides, such as bichloride of mercury, or carbolic acid. Clothing and bedding should be disinfected with superheated steam, sulphurous acid or chlorine. Infected articles of this kind should not be sent to a laundry until they have been thoroughly disinfected by one of the above mentioned means.

In the very beginning of an epidemic, prompt isolation of the sick and thorough disinfection of the surroundings may check its spread, but much cannot be expected from these measures, unless the local sanitary conditions are such as to offer a hindrance to the development of the cholera poison. It is plain, therefore, that prophylactic measures against cholera, if they shall be effective, must be brought into requisition before the epidemic has begun. It is of the highest importance that preventive measures be enforced in time.

RELAPSING FEVER.

Relapsing fever was first clearly described by Dr. John Rutty in his 'Chronological History of the Weather, Seasons and Diseases of Dublin from 1725 to 1765.'* Near the end of the last, and in the first half of the present centuries relapsing fever was frequently met with in an epidemic form in Ireland and Scotland. In 1847 the disease invaded a number of the larger cities of England. From 1868 to 1873 it prevailed extensively in England and Scotland. On the continent of Europe it was first observed in Russia in 1833. In Germany it was not recognised as a distinct disease until 1847, but did not prevail epidemically until 1868. Since then it has often been observed in that country.

Relapsing fever is very prevalent in India, where it was first observed in 1856 by Sutherland. In China and in the countries of Africa bordering on the Red sea, the disease has been recognised by observers.

In the United States it was first observed among emigrants in Philadelphia in 1844, and again in 1869. It was conveyed to other places, but has never prevailed extensively in this country. It has not been observed in the United States since 1871.

The predisposing causes of relapsing fever are above all bad sanitary surroundings. Want and overcrowding seem to be much less important factors than in typhus fever.

Although relapsing fever has, since it was first clearly distinguished from typhus and other continued fevers, been recognised as an eminently contagious and infectious disease, it was not until 1873 that its immediate cause became known. In that year Obermeier discovered in the blood of patients ill with this disease a minute, spiral, mobile organism, now known as the *spirillum* or *spirochæte Obermeieri*.

---

* London, 1770.

Obermeier and other observers, prominent among whom is Dr. Henry V. Carter, have demonstrated the constant presence of these organisms in the blood during the attack. Carter and Koch have induced the disease in monkeys by inoculation of the parasite, and Moschutkowski has successfully inoculated it in the human subject. No doubt can exist at the present day that the spirillum of Obermeier is the true cause of relapsing fever.

The preventive measures consist in attention to details of personal hygiene; in other words, local sanitation, disinfection of infected materials, (fomites,) and complete isolation of the sick.

TYPHOID FEVER.

The first accurate clinical accounts of typhoid fever date from the seventeenth century, when Baglivi, Willis, Sydenham and others described cases of fever which in their clinical characters correspond to the disease now known as typhoid fever. Strother, however, in 1729, first gave a description of the anatomical characters of the disease which he says is 'a symptomatical fever, arising from an inflammation, or an ulcer, fixed on some of the bowels.' Bretonneau and Louis in France, Hildenbrand in Germany, William Jenner in England, and Drs. Gerhard and Pennock in this country clearly pointed out the essential distinction between typhoid and other fevers, during the first half of the present century.

At the present day, typhoid fever is met with everywhere throughout the world. It is at nearly all times a constituent of mortality tables. It affects by preference persons between the ages of fifteen and thirty years, although no age is entirely exempt. It is always more prevalent in the autumn and winter.

The disease is probably due to an organic poison, which gains entrance into the body through the respiratory or digestive tract. Recent observations of Klebs and Eberth seem to indicate that the morbific agent is a micro-organism termed the *bacillus typhoideus*. The exact relation of this organism to the disease has not been clearly worked out. It is found in the intestinal canal, and especially in the characteristic intestinal lesions of this fever. The infective agent is probably contained in the dejections of patients, but does not become active until after undergoing some change outside of the body. The disease is not immediately contagious like typhus fever.

The medium through which the poison is introduced into the body may be drinking water, food, milk or other articles containing the infective agent. Localised epidemics due to infected water or milk have been frequently reported.*

The typhoid poison is supposed to be developed in cess-pools, sewers, and soil polluted by the products of animal decomposition. Whether it ever originates *de novo* in such places is a much disputed proposition. At present the evidence is in favor of the view that cases of typhoid fever are always derived from pre-existing cases. The germ may develope in sewers and be carried in the sewer air from place to place; it may be carried into the soil from cess-pools receiving typhoid dejections and there undergoing development, may ascend through houses with the ground air, or may drain into wells and pollute the drinking water. By the admixture of such water with milk, or other food, the disease may be propagated. It is also believed that the effluvia from typhoid discharges may be absorbed by water or milk, and thus infect these articles.

---

* See ante, p. 45-47.

The prophylactic measures against typhoid fever comprise isolation of the sick, prompt disinfection of the discharges, and cleanliness in its widest sense. The water and food supplies must be carefully guarded against contamination with the poison, and all decomposing animal matter, and excreta must be removed from the immediate vicinity of dwellings. The requisites for prevention may be summed up as pure air, pure water, uncontaminated food, and a clean soil.

TYPHUS FEVER.

Wide-spread pestilences are nearly always accompaniments of famine and war. Of all pestilential diseases, none is so regular in its co-incidence with these conditions as typhus fever. The earliest accounts which unquestionably refer to this disease date from the eleventh century, when it was observed at a number of places in Italy. In the succeeding centuries, isolated accounts of it appeared in the chronicles of the times, but no scientific description of it appeared until the sixteenth century. During the seventeenth, eighteenth, and the early part of the nineteenth centuries, it prevailed extensively throughout Europe. The constant wars and consequent disturbances of the social relations of the people, famines, overcrowding, filth, excesses of all kinds contributed largely to the development and spread of typhus fever. For a number of years past no extensive epidemic of the disease has been observed, although both in this country and in Europe localised outbreaks are frequently met with.

Typhus fever is somewhat more prevalent in the winter and early spring months than during the rest of the year, but not very markedly so.

At present, typhus fever is nearly always limited to times and places where the conditions favoring its development exist. Wherever overcrowding, in connexion with filth, insufficient food and bad habits are present, typhus fever is likely to be a visitor. Thus, in overcrowded and ill-ventilated emigrant ships, in jails and workhouses, and in camps, especially when stress of weather compels the crowding together of soldiers in close huts or barracks, the disease frequently breaks out.

When typhus fever appears in a community, those classes of the people who are subjected to the conditions just mentioned are almost exclusively attacked. In cities, the dwellers in crowded tenements, or in courts and alleys, suffer most severely—are in fact almost the only ones attacked. An exception must, however, be made in the case of hospital physicians and attendants where typhus fever patients are treated. The mortality among these is always large.

Typhus fever is contagious and infectious. An exposure for a length of time to an atmosphere impregnated with the poison may suffice to induce an attack. The poison may also be conveyed from place to place in fomites. Physicians may carry it in their clothing, if they have been exposed to a typhus atmosphere.

The prevention of typhus fever consists in the institution of such measures as will secure pure air, pure water, a clean soil and dwellings, and cleanliness of body and clothing. When an outbreak occurs, the sick should be promptly isolated, the well persons removed from the building in which the cases have occurred, and efficient measures of disinfection carried out. The sick should be treated in the open air as much as possible.

YELLOW FEVER.

The West India islands, the Gulf coast of Mexico, the northern part of the Atlantic coast of South America, and a limited section of the west coast of

Africa constitute the present home of yellow fever. From this area (the so-called 'yellow-fever zone') the disease is frequently transported to contiguous or distant countries. The south Atlantic and Gulf coasts of the United States, and the shores of the Caribbean Sea are most liable to the epidemic visitation of this pestilence.

The first trustworthy account of an epidemic of yellow fever dates from the year 1635, when it prevailed on the island of Guadeloupe. This and the adjoining islands of Dominica, Martinique and Barbadoes were invaded a number of times in the fifty years following the above date. Jamaica was invaded in 1655 and Domingo the year after. In 1693 the first appearance of the disease is mentioned in the United States, being observed in Boston, Philadelphia and Charleston. In 1699 it appeared as an epidemic in Vera Cruz, and reappeared in Philadelphia and Charleston. Since the year 1700, the disease has appeared in an epidemic form at one or more places within the present limits of the United States, seventy-nine times, the last considerable invasion being at Brownsville, Texas, and Pensacola, Florida, in 1882.

In South America, yellow fever appeared for the first time in 1740. In 1849, the disease was imported into Brazil, and has since then been endemic. Peru and the Argentine Republic have also suffered several severe visitations of yellow fever since 1854.

On the west coast of Africa, yellow fever seems to be endemic in the peninsula of Sierra Leone, where it has been frequently observed since 1816. It has also prevailed epidemically in Senegambia, and a number of the islands off the northern portion of the west African coast. In Europe, Spain and Portugal have been the only countries to suffer from yellow fever epidemics.

Although the causes of yellow fever cannot be definitely stated, it is well-known that it only occurs endemically within the tropics, and prevails epidemically elsewhere only during the summer. Of 180 epidemics observed in the United States and Bermudas, 154 began in July, August and September. Of the remaining 26, none began in the six months from November to April.

A temperature of 80° F., and a high humidity are generally considered essential to produce an outbreak of the disease. Of other necessary meteorological conditions nothing is known.

It seems to be well-established that the most filthy and insanitary portions of cities are those principally ravaged by yellow fever. The author is convinced from personal observation in Savannah, Memphis and New Orleans that filth is one of the principal factors in the spread of yellow fever. This opinion is also forcibly expressed by many of the most eminent authorities upon the subject.

Yellow fever is not endemic within the limits of the United States, and has probably never originated here. The instances in which it has appeared to do so may be explained by the persistence of the morbific agent through one or more winters, or by a new importation which has escaped observation.

Yellow fever frequently breaks out on shipboard and causes much loss of life. There is no evidence that it originates on ships; it is only acquired after intercourse with an infected ship or an infected place.

The question of the personal contagion of yellow fever has been decided negatively. The disease is infectious and its cause may be transported in fomites, but persons sick with the disease do not communicate it. An infected atmosphere, or one favorable to the poison, is necessary to the propagation of the disease.

The preventive measures indicated against yellow fever appear from the foregoing : they are strict sanitary inspection to prevent the introduction of a person sick with the disease ; to prevent the introduction of clothing or other fomites from a suspected locality without thorough disinfection, and such a condition of public and private sanitation as will prevent the development of the poison, should the latter, perchance, be introduced.

When the disease becomes epidemic in a city, the inhabitants should be removed to temporary camps beyond the infected area. The experience of the city of Memphis in 1879, encourages the hope that by prompt depopulation of cities, and strict enforcement of sanitary measures in the camps, the terrors of yellow fever can be largely averted. The sick should be promptly isolated, and no one except attendants permitted to have intercourse with them.

SCARLET FEVER AND MEASLES.

The early history of these two contagious eruptive fevers is inextricably blended together. Up to the latter half of the seventeenth century, the distinction between the two was not made by writers. Sydenham was among the first who clearly separated scarlet fever from measles, and gave it a distinct name. Since the great English Hippocrates, the essential character of scarlet fever has been recognised by all physicians, and it is now never, or but rarely, confounded with measles.

Of the two diseases, measles is somewhat more generally prevalent, although both occur usually in epidemics. There is hardly a country in which measles has not been observed, while the continents of Asia and Africa have remained measurably exempt from scarlet fever up to the present time, although epidemics have been recorded in India and Japan.

Hirsch states that scarlet fever was first observed in this country in 1735, at Kingston, Massachusetts, quoting as authorities Dr. Douglass, of Boston, and Dr. Colden, of New York. The latter, however, in a letter to Dr. Fothergill,* clearly describes diphtheria, and not scarlet fever. Its distribution is now general, but it is said to be much milder in the southern, than in other portions of the United States. The prevalence of measles is not limited to any geographical section.

Epidemics of measles usually begin during cold weather. Of 530 epidemics observed in Europe and North America, 339 occurred during the colder, and 191 during the warmer months. In 213 of these, the height of the epidemic occurred 135 times in winter and spring, and only 78 times during summer and autumn. Scarlet fever epidemics occur more frequently in autumn than at any other season.

The cause of scarlet fever or of measles is not to be sought in climatic influences, insanitary surroundings, or special natural conditions of air, water or soil. Both diseases are contagious and infectious, and the contagion is transmitted either by fomites, (clothing, letters, etc.) infected air, drinking-water, or milk.

The measures for the prevention of both diseases are isolation and thorough disinfection.

DIPHTHERIA.

Under the names of Syriac and Egyptian ulcers, Aretæus, a writer of the second century, described various forms of malignant sore throat. The disease now called diphtheria prevailed at various places in Europe during the middle ages. In this country it was first observed about the middle of the last century, and in 1771, Dr. Samuel Bard, of New York, described it very accurately. Although repeated severe outbreaks

* Medical Observations and Inquiries, London, 1776, Vol. I., p. 211.

occurred in Europe in the early part of the present
century, it was not until 1857 that it again attracted
attention by its epidemic prevalence in the United
States. Since that time it has spread throughout the
country, and is at present one of the most generally
diffused, as well as one of the most fatal of the contagious diseases. In certain epidemics its malignancy
is very marked, while in others it seems to be rather
a mild affection.

Diphtheria is personally contagious; it may be
transmitted by inoculation, as well as by inhaling an
infected atmosphere. The virus is supposed by some
investigators to dwell in a micro-organism, whose virulence is dependent upon peculiarities of environment.*

The question as to the identity of diphtheria and
croup is not merely a clinical one, but has an important bearing upon preventive medicine. If croup is a
non-contagious and non-infectious disease, as is held
by many, no precautions will be necessary to prevent
its spread to healthy persons. If, on the other hand,
diphtheria and croup are identical in nature, the danger of infection is equally great in both diseases. Inasmuch as it is frequently impossible to positively
decide upon a diagnosis, it would be well to consider
the identity of the two diseases as established, and act,
so far as preventive measures are concerned, as if all
were cases of diphtheria.

Diphtheria is inoculable upon animals, and may
through this medium be transmitted to man.

Persons sick with diphtheria should be carefully
isolated ; no one but the immediate attendants should
be allowed to come in contact with the patients. Table
utensils, bedding and clothing used by the sick should
be thoroughly disinfected by steam or boiling water
before being used by others. Intimate contact with

---

* Memoir on Diphtheria, by H. C. Wood and H. F. Formad, Report National
Board of Health for 1882.

the sick, such as kissing, should be strictly prohibited. After death or recovery of the patient, the apartment occupied during the illness should be disinfected with chlorine or sulphurous acid gas.

Children recovering from diphtheria, scarlet fever, measles, or small-pox, should not be permitted to attend school for at least four weeks after recovery. It is believed that there is danger of infection for a period about as long as this, and besides, the patients are apt to be weakened from the effects of the disease, and not able to resist the strain of continuous mental effort.

DENGUE.

The disease known as break-bone fever, dandy fever and by various other names, was first observed in the United States in 1780 by Dr. Benjamin Rush. Dr. Rush describes an epidemic which prevailed during the summer and early autumn of that year, under the name of 'bilious remittent fever,' but the symptoms of the disease hardly leave any doubt that it was dengue. In 1779 and 1780 it was also observed on the Coromandel coast, in Egypt and on the island of Java. In 1784 to 1788 dengue also prevailed in various cities of Spain. In 1818 an epidemic appeared in Lima, in which nearly every one of the 70,000 inhabitants was attacked.

In 1824-5 the disease again prevailed widely in India where it was known as the 'three days fever.' Isolated outbreaks occurred in that country until 1853 when it again appeared as a wide-spread epidemic, and in 1872 another epidemic outbreak occurred in the east extending from eastern Africa to Arabia, India and China.

In 1826, an epidemic of dengue appeared in Savannah, and in the following two years spread over the southern portion of the United States and the West Indies, reaching the northern coast of South America.

In 1845 to 1849 the disease was observed in Rio Janeiro; in 1848-50 in the south Atlantic and Gulf states. In 1854 it was observed in southern Alabama, and in the same year in the West Indies. In 1873 another epidemic appeared in the lower Mississippi valley, and in 1880 an outbreak of some extent occurred in New Orleans, Charleston and other places on the Gulf and south Atlantic coasts.

Dengue always begins in the summer or early autumn, and ceases abruptly with the advent of cold weather. It is almost exclusively limited to hot countries. It spreads with extreme rapidity wherever it appears. It is not contagious; the manner of its propagation is not known. The susceptibility to the disease appears to be almost universal; it frequently prostrates the majority of the inhabitants where an outbreak occurs. During the epidemic in Calcutta in 1871-2, seventy-five per cent. of the population were attacked. In the United States similar epidemics have been repeatedly observed.

Dengue is rarely fatal. It seems to be propagated through the atmosphere. No measures of prevention are known or available.

EPIDEMIC INFLUENZA.

Accounts of epidemic influenza can be traced back to the year 1173, when the disease was observed coincidently in Italy, Germany and England. It has prevailed epidemically, at varying intervals, to the present time. In the fourteenth century, three epidemics are recorded; in the fifteenth, four; in the sixteenth, seven; in the seventeenth, forty-six. Of these, fifteen were very extensive, some of them prevailing over both hemispheres contemporaneously.

On the American continent, influenza was first recorded in 1627, when it prevailed in New England, where it again broke out in 1655. Following this there is no notice of the disease in America until 1732, when an epidemic began in the New England States, which extended over the entire globe. Epidemics occurred during the remainder of the eighteenth century in 1737, 1757, 1761, 1767, 1772, 1781, 1789 and 1798. During the present century the disease has prevailed more or less extensively in this country at thirteen different times, the last epidemic of any considerable extent being in 1879.

A curious feature of epidemics of influenza is the coincident occurrence of outbreaks of a somewhat similar affection among animals, horses and dogs being especially attacked.

Influenza is an acute, specific, infectious disease, not directly contagious. The infection is apparently produced or transmitted in the air. The disease frequently appears over a large area of country almost simultaneously. Peculiarities of climate, season, meteorological conditions, geological formation, or racial characteristics, have no apparent influence upon the causation or spread of the disease. It occurs more frequently in the winter and spring than during the summer or autumnal months. The investigation into the epidemic of influenza among horses, referred to in a previous chapter,* seems to indicate, however, that a moist and impure atmosphere intensifies the disease.

No measures of prophylaxis can be indicated, except avoidance of anything tending to depress the vital powers.

EPIDEMIC CEREBRO-SPINAL MENINGITIS.

This disease was first recognised in Geneva in 1805. In the following year it was noted in various places in the United States. Both in Europe and this country,

---

* Chapter I., p. 15.

localised outbreaks of the disease occurred between the dates above mentioned and 1816. At this time the disease seemed to die out altogether, but in 1822 it reappeared in various parts of Europe and America.

Cerebro-spinal meningitis appeared in 1857 in the south-west of France, and during the following ten years spread over a large part of the country. Algiers, Italy, Denmark and Ireland were also visited by the scourge. In 1854 and 1861, Sweden experienced its ravages, and in 1859, Norway was invaded by the disease, which continued for nearly a decennium in the latter country. From 1860 to 1867 the disease prevailed in Holland, Portugal, Germany, Ireland and Russia.

After the termination of what may be called the first epidemic, in 1816, cerebro-spinal meningitis was not again observed in this country until 1842. In the eight years succeeding, it prevailed epidemically throughout almost the whole United States. From 1861 to 1873, it was noted frequently in various parts of the country. Since the latter year the reports of its occurrence in this country have been limited to sporadic cases or localised outbreaks.

Cerebro-spinal meningitis is an acute infectious disease, very fatal in its tendency. It is probably not personally contagious. Climate has no influence upon its origin, but season seems to stand in a positive relation to its causation. About three-fourths of the epidemics noticed have occurred during the winter and spring months. The disease seems to show no preference for peculiarities of topographical or geographical formation. Overcrowding, overwork, and uncleanliness have an important influence in determining an outbreak. It is especially a disease of youth and adolescence. Out of 975 cases occurring in New York, only 150 were over twenty years of age, while of the remainder, 665 were under ten.

The prophylactic measures to be adopted against cerebro-spinal meningitis consist in careful attention to the sanitary conditions of dwellings and streets, avoidance of overwork and overcrowding during times of epidemic, isolation of the sick, and disinfection of the sick-room after the termination of the disease.

SYPHILIS.

In the year 1494 Charles VIII, of France, in command of a large army invaded Italy, and early in the following year besieged Naples. During the investment of the city, a very severe disease, characterised by ulcers of the genitals, violent pains in the head and limbs, and generalised cutaneous eruptions broke out among the besiegers and spread rapidly throughout the army and the civil population. On the return of the army to France after the termination of the war, the disease rapidly spread throughout Europe, and the literature of the early part of the sixteenth century, both medical and lay, teems with references to it.

From the locality and other circumstances connected with its epidemic appearance, the disease acquired various names. Thus the French called it 'morbus Neapolitanus,' or 'mal d'Italie,' while the Italians termed it 'morbus gallicus,' or 'mala franzos.' At a very early period it was, however, clearly recognised that the disease was communicated during sexual intercourse, and hence it was usually described in medical writings under the name 'lues venerea,' while in the popular literature it still figured as the Frenchman's disease (morbus gallicus). The name *syphilis* was first used in a poem descriptive of the disease written in 1521 by Fracastor, a physician of Verona.

The extraordinary outbreak of the disease toward the end of the fifteenth century, led to many speculations concerning its origin. As it attacked persons in all ranks and conditions of life, 'sparing neither crown nor cross,' in the words of a contemporary poet, the favorite explanation was that meteorological influences had much to do with its causation. Many ascribed it to the malign influence of the stars. The Neapolitans attributed it to the wickedness of their enemies, the French, while the latter laid the blame on the filth and immorality of the Italians. The Spaniards claimed that it had been imported from America by Columbus, whose first expedition returned to Europe in 1493. There are records, however, which prove that the disease already existed in Italy in the latter year. In other parts of Europe the Jews, who had been driven out of Spain by the terrors of the inquisition; were accused of this, as of many other misfortunes which befell the people. When it was definitely established that the disease was communicated almost solely by sexual intercourse, the theory of its transatlantic origin became very popular. It is a characteristic of human nature to refer the origin of troubles resulting from its own vices to some other source, if possible. This theory of the American origin of syphilis is still held by some writers. Within a few years Dr. Joseph Jones, of New Orleans, claims to have found evidences of syphilitic disease in the skulls and other bones from some of the pre-historic Indian mounds in Mississippi. These observations of Dr. Jones have however, not yet been verified by others.

Although the first great epidemic of syphilis is clearly traceable to the period between the years 1493 and 1496, an examination of the older literature reveals many descriptions of disease which can only be ex-plained by assuming them to refer to syphilis. The Old Testament Scriptures contain numerous references to diseases of the genital organs. In most instances these troubles are ascribed to the wrath of God, although in some cases a pretty shrewd hint is given as to the causation of the affections. Finaly* remarks that the Hebrew word translated in all versions of the Bible by 'flesh,' signifies also the virile member. In this light the references in Leviticus XIII.-XV.; Numbers XXV. 1-9; XXXI. 16-18; Deuteronomy IV. 3; Joshua XXII. 17; I. Samuel V. 6, 9, 12; Ps. CVI. 28-30; I. Corinthians X. 8; Ephesians II. 11, and Colossians II. 13, receive a new interpretation. Numerous innuendoes in the Latin classics, and more or less exact descriptions in the medical writings of Greece, Rome, China and India, leave no room for doubt that venereal diseases, and probably among them syphilis, have existed from the earliest times.

At the present day syphilis is the most widely prevalent of all contagious diseases. In 1873 Dr. F. R. Sturgis estimated that in New York one person out of every eighteen suffered from it. This is considered a moderate estimate. Dr. J. Wm. White, of Philadelphia, pronounces the opinion that 'not less than fifty thousand people of all classes in that city are affected with syphilis.' On this basis, Gihon estimates the number of syphilitics in the United States at one time, at two millions.†

The disease is transmitted, in the vast majority of cases, during the performance of the sexual act, but there are numerous other ways in which it may be and frequently is communicated. In the special literature of the subject are records of many cases in which the disease was acquired through a kiss, a bite, the act of suckling (from infant to nurse and conversely), using a pipe, a glass-blowers' mouth-piece, the finger of a midwife, the instruments of the dentist or surgeon, inoculation of syphilitic secretion mixed with saliva in the

---

* Arch. f. Dermat. u. Syphilis, II. Jahrg. 1 Heft. p. 126.

† The Prevention of Venereal Diseases by Legislation. Sanitarian, June, 1882.

process of tattooing and many other ways. Numerous cases have been reported where physicians were inoculated on the finger, while examining a syphilitic patient.

The prophylactic measures which suggest themselves from a consideration of the nature of the disease, are isolation of those infected, regular inspection of the class of persons through whom the disease is most frequently transmitted, *i. e.*, prostitutes, and individual precautions against acquiring it. Greater attention to cleanliness of the genital organs on the part of those indulging in promiscuous intercourse, would aid largely in reducing the number of cases of syphilis.

DISEASES OF ANIMALS COMMUNICABLE TO MAN.

Sheep-pock.—This a highly contagious and infectious disease of sheep, resembling in its symptoms, course and fatality, small-pox as it occurs in the human race. It is believed by Bollinger to be different from the form of small-pox produced in sheep, goats, horses, and other animals by the inoculation of human small-pox. Sheep-pock can be inoculated upon other animals and man but only produces a local disease at the point of inoculation, in the latter. Sheep may be protected against this disease by inoculation with sheep-pock virus (ovination), or by vaccination with vaccine lymph. The peculiarity of sheep vaccinia is that it is a more or less generalised disease, the pustules being distributed over the body. Sheep-pock when inoculated upon human beings does not produce a generalised infectious disease, but remains entirely local.

Actinomycosis.—Veterinarians have frequently observed a disease attacking the jaws of cattle and producing tumors, often with ulcerated surfaces. The bone is usually involved. The disease has heretofore been generally considered a sarcomatous growth. It is not seldom observed among the cattle in the western stockyards, where it is known in the vernacular as 'swell-head.' Recent investigations by Ponfick have shown that the growth consists of a vegetable parasite (actinomyces), and that it is inoculable upon other animals, and may be conveyed to man. A considerable number of cases have been observed in human beings in Germany, where the disease was first described by Ponfick, and very recently two cases have been reported in this country.*

Bovine Tuberculosis.—(Perlsucht). In cattle, tuberculosis occurs in two forms, miliary tubercles and cheesy masses in the lungs, and firm, pearly nodules on the serous membranes. These nodules do not break down, but may become calcified.

Bovine tuberculosis is a frequent disease among cows kept in damp, dark and ill-ventilated stables. The disease, which is essentially the same as human tuberculosis, tubercle bacilli being present in the neoplasms, is believed by many to be transmissible to human beings by means of the milk or flesh of tuberculous animals. The sale of meat of tuberculous cattle should be prohibited.

Rabies.—Hydrophobia in the brute, and its communicability to man through a bite has been known from the remotest antiquity. It occurs in dogs, foxes, wolves, horses and other animals, and may be transmitted from any of them to human beings.

The contagium of rabies, the infective poison, is contained principally in the saliva, and is usually inoculated by the teeth of the mad animal.

Pasteur seems to have demonstrated that the virus of rabies consists in a micro-organism found in the saliva of the rabid animal. By cultivation in appro-

---

* Boston Med. and Surg. Journal, Oct. 16, 1884. p. 377.

priate culture media, the infectiveness of this organism can be modified. When inoculated, after cultivation according to Pasteur's methods, it protects the animal against the effects of rabies virus.*

Anthrax.—Anthrax or splenic fever (milzbrand) is an acute, highly contagious and infectious disease of herbivorous animals which may be transmitted, by inoculation or the ingestion of the virus, to other animals and to man.

The disease is due to a minute vegetable organism which is found in the blood and tissues of the diseased animals. This organism, *bacillus anthracis*, was first discovered by Pollender, and has been thoroughly investigated by Davaine, Pasteur, Koch and others.

Inoculation of these bacilli or their spores always produces the disease in susceptible animals. Skins of animals not infrequently contain the virus which may then gain access to the blood of persons engaged in handling them. Knackers, butchers, wool-sorters, and other persons liable to come in contact with sick animals, or handling their flesh or hides, are subject to the infection, either by direct inoculation (through abrasions of the skin, etc.,) or by inhalation of the spores of the bacillus. An intestinal form of anthrax in man, *mycosis intestinalis*, is sometimes produced by the consumption of meat of animals suffering, when killed, of splenic fever. Numerous instances have been reported. The diagnosis has been verified by discovering the bacillus of anthrax in the blood and various organs of the individuals attacked.

In view of the dangerous character of the disease, persons coming in contact with animals suffering from anthrax should be warned of their peril. In order to protect other animals in a herd, strict isolation of the

*Transactions International Medical Congress of 1884. Med. Record, Aug. 30 1884, p. 247.

infected, thorough disinfection of the stables occupied by them, and deep interment of the cadavers of those dead from the disease are indicated.

Glanders.—Glanders or farcy is a very fatal contagious disease of horses which may be communicated to other animals and to man. The cause of glanders has recently been discovered to be a bacillus resembling the bacillus tuberculosis. Pure cultures of this bacillus were inoculated into animals, and followed by glanders in a number of the cases.

The infection in man may occur either upon the seat of excoriations of the skin or mucous membranes, especially those of the nose, conjunctiva, and possibly by inhalation of infective particles floating in the air.

Animals with glanders should be promptly killed, and their cadavers cremated or deeply buried. No part of the body of any animal dead with glanders should be allowed to be used. Infected stables should be thoroughly disinfected.

[The works of especial value to students who desire fuller information upon the subjects treated in this chapter are the following :

HIRSCH : Handbuch der Historisch-Geographischen Pathologie, 2te Aufl., Stuttgart, 1883. HECKER : The Black Death. Transl. by B. G. Babington. DEFOE : Journal of the Plague in London ; ROHLFS : Die Orientalische Pest. MARSON : Small-pox, in REYNOLDS' System of Medicine, Vol. I. SEATON : Vaccination, *ibid*. TROUSSEAU : Clinical Medicine, Vol. I. Fifth Annual Report Illinois State Board of Health. WOODWORTH and MCCLELLAN : Cholera Epidemic in the U. S. in 1873. CHAILLÉ : Report of Yellow Fever Commission, in Annual Rep't Nat'l Board of Health for 1880. WOOD and FORMAD : Memoir on the Nature of Diphtheria ; ibid, 1882. THOMPSON : Annals of Influenza. STILLÉ : Epidemic Meningitis. MUELLER : Die Venerischen Krankheiten im Alterthum. LANCEREAUX : Traité de la Syphilis. BOLLINGER : Ueber Menschen u. Thierpocken, etc. Samml. Klin. Vortr. No. 116. PONFICK : Die Actinomycose des Menschen. GAMGEE : Hydrophobia and Glanders in, REYNOLDS' System of Medicine, Vol. I. BOLLINGER : Anthrax, in ZIEMSSEN's Cyclopædia, Vol. III.]

CHAPTER XX.

ANTISEPTICS AND DISINFECTANTS.

Much confusion exists in the popular mind and even among physicians, as to the exact meaning of these two terms. By many they are used synonymously, and hence frequently give rise to ambiguity and misunderstanding.

Antiseptics may be defined as substances which retard or prevent fermentation, decay or putrefaction. They interfere with the development of the organisms which are the causes of these processes. They may also retard the development of the organisms which cause infectious diseases. On the other hand, the action of disinfectants is destructive ; they destroy the causes of these diseases. If all infectious diseases are assumed to be due to micro-organisms or germs, disinfectants may be considered as equivalent to germicides. The difference of action of antiseptics and disinfectants can be best understood by assuming that it is merely a difference of degree, not of kind. Disinfectants are generally antiseptics also, but antiseptics are not in all cases disinfectants.

COMPARISON OF ANTISEPTICS AND DISINFECTANTS.

This can best be illustrated by quoting some experiments bearing upon this point. In a recent paper,* Dr. G. M. Sternberg gives the following table, showing the relative disinfectant (destructive of vitality) and antiseptic (preventive of development) power, the micrococcus of pus being the organism experimented with :

---

* American Journ. Med. Sciences, April, 1883, p. 335.

TABLE I.

Agent used.	Percentage used to destroy vitality.	Percentage capable preventing development.
Bichloride of Mercury - -	0.005	0.003
Iodine · · - - -	0.2	0.025
Sulphuric Acid · · -	0.25	0.12
Carbolic Acid · - ·	0.8	0.2
Salicylic Acid and Biborate of Sodium	4	0.5
Alcohol · · - · -	40	10
Sulphate of Iron - -	Failed in saturated solution	0.5
Boracic Acid - ·	"	0.5
Biborate of Sodium - -	"	0.5

It is readily seen, that while all of the substances mentioned acted efficiently as antiseptics in reasonably dilute solutions, only two-thirds were efficient disinfectants or germicides. In order to destroy the vitality of the micrococcus, the solutions were required to be from four to eight times stronger than was necessary to merely arrest its development. In another table the same author gives the results of a larger list of substances, showing the relative germicidal power of the agents experimented with. The table is as follows :

TABLE II.

	Efficient in the proportion of one part in
Bichloride of mercury (0.005 p. ct.) - - -	20,000
Permanganate of potassium (0.12 p. ct.) - - -	833
Iodine (0.2 p. ct.) · · · · ·	500
Creasote (0.5 p. ct.) · · · · ·	200
Sulphuric acid (0.5 p. ct.) · · · · ·	200
Carbolic acid (1 p ct.) - · · · ·	100
Hydrochloric acid (1 p. ct.) · · · ·	100
Chloride of zinc (2 p. ct) - - - ·	50
Tinct. chloride of iron (4 p. ct.) · · · ·	25
Salicylic acid dissolved by borate of soda (4 p. ct.) · ·	25
Caustic potash (10 p. ct.) · · · · ·	10
Citric acid (12 p. ct.) · · · · ·	8
Hydrate of chloral (20 p. ct.) · · · ·	5

Assuming that infectious diseases are caused by micro-organisms, and that these are different from the micro-organisms of ordinary decay or putrefaction, it can be readily understood that the processes of organic decomposition may themselves act as disinfectants. It is known, for example, that when a fermenting liquid putrefies, the organisms of fermentation disappear and

give place to the organisms of putrefaction (bacterium
termo, etc.) So likewise the bacilli of anthrax and of
tuberculosis are killed by the putrefactive process, if
this takes place in the absence of free oxygen. Furthermore, the reproduction of organisms of a certain
kind ceases when certain chemical (?) changes take
place in their environment. Fermentation in a saccharine liquid ceases and the ferment-organisms die when
the accumulation of the product of the fermentation
(alcohol) has reached a certain proportion, although
there may still be undecomposed sugar present. In
like manner it is intelligible that the products of micro-
organisms may eventually destroy their producers, and
so place a limit to the morbid process. The specific
cause of small-pox, yellow fever, cholera, and similar
infectious diseases is rapidly destroyed when decomposition of the corpses of those dead with such diseases
sets in. Hence, the reason why infectious diseases are
not spread from cemeteries.

These facts also emphasise the necessity of making
any attempted disinfection thorough ; in other words,
the disinfectant used should be capable of absolutely
destroying all disease germs. The addition of an antiseptic to material containing disease germs may merely
preserve the latter by retarding their development.
By referring to table I., it will be seen that sulphate of
iron, which is so frequently used to disinfect typhoid
discharges, sinks and cess-pools, is not a germicide at
all, while carbolic acid can only be depended upon if
used in the strength of one part in one hundred.

From the foregoing it may be gathered that disinfection consists chiefly in a struggle against organised
disease germs.* As, however, experiments and observations have shown that the life-history of disease-
germs varies with the different organisms involved, it

---

* MUELLER UND FALK, in Realencyclopædie d. ges. Heilk, Bd. IV., p. 62.

becomes evident that specific directions concerning disinfection can be given only when the life-history of the
specific organism is known. A valuable contribution
to this knowledge has been made by Sternberg,* who
experimented with micrococcus of pus, of septicemia
of rabbits, bacterium termo, and the bacteria of broken-
down beef-tea., These experiments showed that as a
germicide to all these organisms the bichloride of mercury stands first in the list. In the proportion of
1 : 20,000 it destroyed the two forms of micrococci
mentioned, while the bacteria required twice this
strength, or 1 : 10,000. Iodine was effective in a 1 : 500
solution on all the organisms, while carbolic acid was
effective only on the first three when the strength of
the solution was 1 : 200, or 1 : 100. The bacteria of
broken-down beef-tea failed, except in one instance, to
yield their vitality to a four per cent. solution.

According to Koch, the following solutions destroy
the spores of anthrax bacilli within twenty-four hours:

Chlorine water.
Bromine, 1 : 50.
Iodine water.
Osmic acid, 1 : 100.
Permanganate of potassium, 1 : 20.
Bichloride of mercury, 1 : 20,000, in ten minutes.

These results agree, in general, with those obtained
by Sternberg and other observers, and it may be accepted that corrosive sublimate, iodine, bromine, chlorine and permanganate of potassium in solution in
the proportions given are trustworthy disinfectants or
germicides. As antiseptics they are efficient in much
more dilute solutions, as shown in table I.

Liquid disinfectants may be used to destroy the
infectiveness of excremental materials, to wash infected clothing, furniture, rooms or vessels, and for dress-
ing wounds or ulcers and rendering them aseptic.

---

* Op. cit., . 331.

The principal gaseous disinfectants are sulphurous acid, chlorine and nitrous acid. The most generally used is sulphurous acid. Dr. Sternberg has shown* that exposure to an atmosphere containing one per cent. of sulphurous acid gas for at least six hours will destroy the infectiveness of vaccine virus. The same results were obtained with chlorine and nitrous acid. Sulphurous acid is, however, more readily applied than the others mentioned. For an ordinary-sized room, one pound of roll sulphur burned in the air of the room and confined in it by closing doors and windows, should prove an efficient disinfectant. For greater safety, twice this quantity should be used. The room should remain closed for ten or twelve hours, and then all windows opened and the air allowed free access. Every room should be thus disinfected after it has been occupied by a case of scarlet fever, measles, small-pox, typhoid, typhus or yellow fever, diphtheria, cholera or similar diseases. Gaseous disinfectants act with greater power when the air is moist than when it is dry. It would, therefore, be good practice to render the air of any room to be disinfected, moist, by means of steam or a spray apparatus before using the gaseous disinfectant. The spray itself might be rendered disinfectant by using a solution of bichloride of mercury, carbolic acid, or chlorine water.

Dry or moist heat (hot air or steam) are the best disinfectants where they can be applied. A temperature of from 150° to 300° Fahr., is probably sufficient to destroy the organic cause of any disease. Steam is more efficient than dry heat. While the above temperatures will generally destroy disease-germs, it has been shown by Tyndall† that the desiccated spores of the common hay bacillus (bacillus subtilis) may withstand

---

* Report National Board of Health for 1880, p. 318-322.

† Essays on the Floating Matter of the Air.

a higher temperature. When steam or hot air (temp. 230°–250° F.) are used as disinfectants they should be under a pressure of one-fourth to one-half atmosphere, in order that the material to be disinfected may be entirely permeated by the heat. When steam is employed, the steaming should continue at least an hour. Dry heat should be continued from one to three hours.

The following substances are antiseptics, but in the strength given cannot be depended upon as disinfectants:

TABLE III.

Thymol,	1 : 80,000.
Bichloride of mercury,	1 : 40,000.
Oil of mustard,	1 : 33,000.
Acetate of alumina,	1 : 6,310.
Bromine,	1 : 5,597.
Picric acid,	1 : 5,000.
Iodine,	1 : 4,000.
Sulphuric acid,	1 : 800–1 : 3,353.
Permanganate of potassium,	1 : 3,000.
Camphor,	1 : 2,500.
Eucalyptol,	1 : 2,500.
Chromic acid,	1 : 2,200.
Chloride of aluminum,	1 : 2,000.
Hydrochloric acid,	1 : 1,700.
Benzoic acid,	1 : 1,439.
Quinine,	1 : 1,000.
Boric acid,	1 : 200–1 : 800.
Salicylic acid,	1 : 200–1 : 800.
Carbolic acid,	1 : 500.
Sulphate of copper,	1 : 400.
Nitric acid,	1 : 400.
Biborate of soda,	1 : 200.
Sulphate of iron,	1 : 200.
Creasote,	1 : 200.
Arsenious acid,	1 : 100.
Pyrogallic acid,	1 : 62.
Tr. chloride of iron,	1 : 25.
Alcohol,	40 to 95 per cent.

[The following additional works are recommended for study in connexion with this chapter:

STERNBERG AND MAGNAN: The Bacteria, Second Edition. FLUEGGE: Fermente und Mikroparasiten,.in VON PETTENKOFER UND ZIEMSSEN'S Handb. d. Hygiene. 1 Th., 2 Abth., 1 Hft. WERNICH: Desinfectionslehre zum praktischen Gebrauch.]

CHAPTER XXI.

QUARANTINE.

THE word quarantine is derived from the Italian *quarantina*, meaning forty, and was originally applied to the period of detention of persons arriving at a port or city from a place where infectious diseases prevailed. The first quarantine regulations date from the fourteenth century, when certain measures against the introduction of the plague were enforced in the city of Florence. The period of detention during which the infected or suspected individuals were prohibited from holding intercourse with the inhabitants of the city was forty days, whence the name quarantine. In the succeeding century the authorities of Venice, then at the climax of her power, instituted a system of quarantine, which formed the model of similar establishments until recent times. The penalties for infractions of the quarantine regulations were exceptionally severe.

At the present day the word quarantine has lost its original significance. The ancient quarantine of detention has given place, theoretically at least, to a quarantine of observation. The old notions still keep a strong hold on the official mind however, and too much faith is still placed on the protective influence of a quarantine of detention.

The object of quarantine is to prevent the introduction into a city, state, or country, of contagious or infectious diseases. Obviously, the most certain way to effect this would be to absolutely prohibit all communion with persons and things from infected countries. But the demands of modern commerce and the necessities of international intercourse clearly make this impracticable. No civilised community can afford to isolate itself from the rest of the world for any considerable length of time. Measures designed to prevent the introduction of contagious diseases must be of a character not to interfere too greatly with travel and trade.

The failure of the old system of detention to prevent the entrance of infectious diseases has caused some countries to abolish quarantine altogether, under the conviction that no quarantine system could effect the desired object. This is, however, a one-sided view of the matter. Quarantines can be made effectual safeguards, if carried out in accordance with the modern knowledge of the diseases to be kept out. The modern notion of quarantine comprehends much more than the old. The following points illustrate its principal features. The inspecting officer must know:

1.—The sanitary condition of the port of departure.

2.—The original places of departure of passengers, crew, and cargo, and their sanitary conditions.

3.—The sanitary condition of the vessel at the time of leaving port.

4.—The sanitary history of the vessel, crew and passengers during the voyage.

5.—The sanitary condition of the vessel and contents at the time of arrival and inspection.

The first, second and third conditions should be fully given in the consular bill of health. In order that this may be trustworthy, a competent medical officer should be attached to every consulate at places of embarkation or shipment. The duty of this medical officer should not be merely to inspect the ship and satisfy himself of her cleanliness before departure, but he should be required to keep himself informed of the sanitary history of all places whence passengers and their effects, or cargo come, in order that on his advice a clean bill of health may be refused to any vessel, which fails to come up to the required conditions. The medical attaché might also be required to inspect the quarters and provisions of crews and passengers, in order to secure compliance with the laws in existence regulating these matters. The passengers and crew should be examined and if any are found suffering from contagious diseases they should not be permitted to proceed. No one should be allowed to set foot on a ship who has not been previously vaccinated or protected by an attack of small-pox.

The sanitary condition of the vessel, crew and passengers during the voyage is ascertained from the log which is the official journal of the ship, and from the report of the ship's doctor, if there is such an officer on board. All cases of sickness or deaths which have occurred during the voyage, must be reported to the quarantine officer.

The present condition of the vessel is ascertained by actual inspection at the quarantine station. Through this inspection the health of the crew and passengers, and the condition of the ship, as regards cleanliness, character and state of cargo, etc., are determined. With the knowledge obtained from these various sources, the quarantine officer can now act intelligently.

If the sanitary history of the vessel, passengers, crew and cargo is good ; that is to say, if the vessel is clean, the passengers and crew healthy on embarkation, the cargo and passengers from a non-infected place, the port of departure healthy, and if no contagious disease appeared upon the vessel during the voyage, or is present at the time of inspection, the ship may be given 'free pratique ;' i. e. may be allowed to enter the harbor, discharge her cargo and passengers without danger of communicating infectious disease.

If, however, the passengers and cargo come from a locality where a contagious disease is epidemic at the time of departure, or if such an epidemic exists at the port of embarkation and shipment, or has broken out during the voyage, the vessel must be declared infected, no matter whether she has a clean bill of health or not. The passengers and crew with their baggage are disembarked, the latter disinfected, and the former detained on shore until the vessel has been disinfected. All sick are removed to the quarantine hospital until the sickness comes to an end. After the ship and cargo have been thoroughly disinfected, for which purpose it may be necessary to unload the latter, the ship may be allowed to proceed to her wharf and discharge, or the cargo may be directly transshipped at the quarantine station. The passengers and crew must be detained at quarantine until the period of incubation of the disease quarantined against has passed, counting from the time of departure from the infected port, or from the appearance of the last case among them. Hence, if the vessel came from a yellow fever or cholera port, the period of detention need not exceed five days from time of departure, or the last exposure.

Typhus fever may have a period of incubation lasting twelve days, and in small-pox, the usual period is two weeks, hence when these diseases are quarantined against, the detention of passengers must cover the time of these respective periods from the date of the last exposure.

The proper equipment of a quarantine station comprises residences for the quarantine officers and employés, hospitals for the treatment of the sick, barracks to lodge passengers, boarding boats, wharves to secure vessels, warehouses in which freight and baggage can be stored, disinfecting chambers, and apparatus by means of which disinfectants can be forced into vessels and disinfecting chambers.

A quarantine carried out in accordance with these principles might appropriately be called a quarantine of observation and disinfection, with incidental detention.

[More detailed information on quarantine may be found in the following works:

VANDERPOEL: Quarantine, in BUCK's Hygiene and Public Health, Vol. II. General Principles affecting the Organisation and Practice of Quarantine. Public Health. Vol. I. PERRY: Effectual External Sanitary Regulations Without delay to Commerce, *Ibid.* BAKER: International Rules of Quarantine, *Ibid.* Vol. V.]

CHAPTER XXII.

VITAL STATISTICS.

The registration of vital statistics comprises the recording of the births, marriages, deaths, and diseases of a city, state, or nation. In no other way can a knowledge of the' health of the inhabitants of such communities be obtained. For smaller, or special communities, such as armies, navies, schools, or special classes of·workmen, the health status may be ascertained by direct methods, but for larger communities this is clearly impracticable, and the sanitarian is obliged to depend upon the census, and the registration of births, marriages and deaths.

From a sanitary point of view, the most important object of a registration of vital statistics is to 'give warning of the undue increase of disease or death presumed to be due to preventable causes, and also to indicate the localities in which sanitary effort is most desirable and most likely to be of use.'*

The duty of registration should devolve upon the sanitary administration. Local and State boards of health would seem to be the most appropriate media for collecting information bearing upon births, diseases and deaths. It would seem also to be most·appropriate to require the attending physicians to make reports of deaths and of cases of contagious disease to the health authorities.

REGISTRATION OF DEATHS.

The data entered upon the record of death should comprise the name, age, sex and color of decedent, nativity, descent, occupation, civil condition, date,

---

* J. S. Billings: Registration of Vital Statistics; Am. Jour. Med. Sciences, Vol. LXXXV, p. 87.

place and cause of death. Under the heading 'descent,' the birthplace of each parent should be given. Occupation should be accurately specified. The place of death should indicate the exact locality, (number of street) where it occurred. Both proximate and predisposing causes of the death should be entered, and any complications which may have influenced the fatal termination should be noted on the record.

This record should be in the possession of the local health authority before a permit for the burial of the deceased is granted. If this is not insisted upon, the report will soon be omitted and the registration become defective.

REGISTRATION OF BIRTHS.

The collection of data for an accurate registration of births is much more difficult than the record of deaths. Instead of requiring physicians and midwives in attendance at the confinement to report births, it would be more equitable, and probably more effectual to compel the parents, under a penalty for failure, to record the birth of each child at the board of health. The items usually included in birth returns are: date and place of birth; sex and color of child; names of father and mother; parents' nativity, and age; and father's occupation. Sometimes the residence of the mother, number of children previously borne by the same mother, whether the child is legitimate or not, and various other details are also added. It is evident that for sanitary purposes most of this information is entirely irrelevant. It seems to the author that for the purpose of the sanitarian and medical statistician, the date and place of birth, sex and color of the child, and age, nativity, and occupation of both parents are sufficient.

REGISTRATION OF MARRIAGES.

The record of marriages is of no interest to the sanitarian. If, however, the registration could be made by a competent medical man, and the physical condition of the contracting parties noted, valuable deductions might be made in time, especially if the parties themselves and their offspring could be kept under observation for many years. This, however, is so manifestly impracticable that it barely deserves notice in this place.

REGISTRATION OF DISEASES.

As has been seen in chapter XIX, a large class of diseases are communicable from one individual to another, either directly, by contact, or mediately, by infection. In large communities it is therefore important that the sanitary authorities should possess information of the presence and prevalence of these diseases, in order that measures may be instituted for their restriction. It is true that in most cases the registration of deaths gives but too mournful evidence of the more fatal of the diseases of this class, but destructive epidemics could probably be frequently averted if preventive measures could be enforced early. Besides, in the case of dengue and epidemic influenza, the death-rate is so small, that if the registration of deaths were alone depended upon, no evidence whatever might be attainable of the epidemic prevalence of such diseases.

The registration of prevailing diseases is, therefore, one of the most important duties of the registrar of vital statistics. Prompt notice of all cases of infectious, miasmatic or contagious diseases coming under their *professional* notice should be required of all physicians. It is unquestionably just, however, that the physicians required to perform this duty should be properly compensated by the public, whose interests they serve.

DEATH-RATE AND BIRTH-RATE.

In order to calculate the annual death-rate. of a place, two facts are required to be known: first, the actual or estimated population (generally obtained from the census), and second, the number of persons who died within the district. The number of deaths is then divided by the population, which gives the death-rate for each individual for the year. . To find the death-rate per thousand, the death-rate as found above is multiplied by 1,000. Thus, the total number of deaths in the city of Baltimore during 1883 was 9,380, and the estimated population 408,520. The death-rate for the year was 22.96 per thousand, obtained as follows :

$$\frac{9,380 \times 1,000}{408,520} = 22.96 \text{ per M.}$$

To calculate the annual death-rate of a place for each thousand of the population from the returns for one week, the weekly population for the place is first ascertained, and then the total number of deaths for the week divided by the weekly population, and the quotient multiplied by 1,000. The following concrete example will render this clear :   .

The exact number of weeks ,in a year is 52.17747. The total population is divided by this figure, giving the weekly population. This gives for Baltimore, assuming the above estimate to be correct, a weekly population of 7,829. For the week ending November 1, 1884, the deaths in that city numbered 148. The annual death-rate per thousand, that is to say, the number of deaths in each thousand of population, if the same ratio were maintained throughout the year, is obtained as follows :

$$\frac{148 \times 1,000}{7,829} = 18.9 \text{ per M. per annum.}$$

Out of the above 148 deaths, 32 were from infectious diseases. To find the annual death-rate per thousand of population for this class of diseases, the same calculation is made, thus:

$$\frac{32 \times 1,000}{7,829} = 4.1 \text{ per M. per annum.}$$

Or if the percentage of deaths of infectious diseases be desired, the procedure would be as follows:

$$\frac{32 \times 100}{148} = 21.6 \text{ per cent. of the total deaths.}$$

Sixty-four of the decedents were under five years of age. The death-rate for this class is found in the same manner, for example:

$$\frac{64 \times 1,000}{7,829} = 8.17 \text{ per M. per annum.}$$

or the percentage of these to the total deaths is found as in the last example.

If it be desired to find the rate of infant mortality, *i. e.*, the proportion of deaths among infants under one year of age to the total number of births for the same period, the following formula may be used. In the record above quoted the decedents under one year of age numbered 37; the total number of births for the same week was 157. Hence

$$\frac{37 \times 1,000}{157} = 235.7 \text{ per thousand births.}$$

or nearly one to four.

Thirty-three of these 148 deaths were of colored persons. The death-rate of these to the total population is found in a similar manner to the above; but if it is desired to ascertain the death-rate of the colored population alone, the weekly colored population must first be obtained and the rate calculated from this by the above formula.

Birth-rates are found in a similar manner. The average age at death is ascertained by adding up the ages of all the decedents, and dividing the sum by the number of deaths.

It will be evident on a little thought that there must be many sources of error in calculations based upon such uncertain data as are derived from the registration of births and deaths as conducted in most cities in this country. Besides, the subject of vital statistics is essentially abstruse and requires no little readiness in mathematics to appreciate its profounder bearings. Hence, in the foregoing chapter no attempt has been made to penetrate beyond the immediate practical aspects of the questions involved.

[To those desiring fuller information upon this subject, the following works are recommended :

CURTIS: Vital Statistics, in BUCK's Hygiene and Public Health. BILLINGS: Registration of Vital Statistics, Am. Journ. Med. Sciences, Vol. LXXXV. OLDENDORFF: Morbilitæts und Mortalitæts-Statistik, in Realencyclopædie d. ges. Heilk. Bd. IX. BILLINGS: Papers on Vital Statistics. Sanitary Engineer, Vol. VIII, IX.]

INDEX.